AF567194

Advances in Space Exploration
COSPAR Symposium Series

VOLUME 10

REMOTE SENSING AND MINERAL EXPLORATION

Advances in Space Exploration
COSPAR Symposium Series

Volume 1 K. A. VAN DER HUCHT and G. S. VAIANA (Editors): New Instrumentation for Space Astronomy

Volume 2 E. A. GODBY and J. OTTERMAN (Editors): The Contribution of Space Observations to Global Food Information Systems

Volume 3 W. A. BAITY and L. E. PETERSON (Editors): X-Ray Astronomy

Volume 4 H.-J. BOLLE (Editor): Remote Sounding of the Atmosphere from Space

Volume 5 W. RIEDLER (Editor): Scientific Ballooning

Volume 6 YASH PAL (Editor): Space and Development

Volume 7 R. COWSIK and R. D. WILLS (Editors): Non-Solar Gamma-Rays

Volume 8 A. P. MITRA (Editor): Low Latitude Aeronomical Processes

Volume 9 V. V. SALOMONSON and P. D. BHAVSAR (Editors): The Contribution of Space Observations to Water Resources Management

Volume 10 W. D. CARTER, L. C. ROWAN and J. F. HUNTINGTON (Editors): Remote Sensing and Mineral Exploration

REMOTE SENSING AND MINERAL EXPLORATION

Proceedings of a Workshop of the Twenty-second Plenary Meeting of COSPAR
Bangalore, India
29 May to 9 June 1979

Edited by
W. D. CARTER and L. C. ROWAN
US Geological Survey, Reston, Virginia, USA
and
J. F. HUNTINGTON
CSIRO, North Rhyde, Australia

Published for
THE COMMITTEE ON SPACE RESEARCH
by
PERGAMON PRESS

OXFORD · NEW YORK · TORONTO · SYDNEY · PARIS · FRANKFURT

U.K.	Pergamon Press Ltd., Headington Hill Hall, Oxford OX3 0BW, England
U.S.A.	Pergamon Press Inc., Maxwell House, Fairview Park, Elmsford, New York 10523, U.S.A.
CANADA	Pergamon of Canada, Suite 104, 150 Consumers Road, Willowdale, Ontario M2J 1P9, Canada
AUSTRALIA	Pergamon Press (Aust.) Pty. Ltd., P.O. Box 544, Potts Point, N.S.W. 2011, Australia
FRANCE	Pergamon Press SARL, 24 rue des Ecoles, 75240 Paris, Cedex 05, France
FEDERAL REPUBLIC OF GERMANY	Pergamon Press GmbH, 6242 Kronberg-Taunus, Pferdstrasse 1, Federal Republic of Germany

First edition 1980

British Library Cataloguing in Publication Data

Symposium on Remote Sensing and Mineral Exploration, *Bangalore, 1979*
Remote sensing and mineral exploration. -
(Advances in space exploration; vol.10).
1. Prospecting   2. Remote sensing
I. Title   II. Carter, W D   III. Rowan, L C
IV. Huntingdon, J F   V. COSPAR, *Plenary Meeting, 22nd, Bangalore, 1979*   VI. Series
622'.1   TN270   79-41355

ISBN 0-08-024438-6

In order to make this volume available as economically and as rapidly as possible the authors' typescripts have been reproduced in their original forms. This method unfortunately has its typographical limitations but it is hoped that they in no way distract the reader.

Printed in Great Britain by A. Wheaton & Co., Ltd., Exeter

Workshop on Remote Sensing and Mineral Exploration

Organized by

THE COMMITTEE ON SPACE RESEARCH — COSPAR

and

THE INDIAN NATIONAL COMMITTEE FOR SPACE RESEARCH — INCOSPAR

THE INDIAN SPACE RESEARCH ORGANIZATION — ISRO

THE INDIAN NATIONAL SCIENCE ACADEMY — INSA

Sponsored by

THE COMMITTEE ON SPACE RESEARCH — COSPAR

THE INTERNATIONAL GEOLOGICAL CORRELATION PROGRAM — IGCP of IUGS and UNESCO

THE GEOLOGICAL SURVEY OF INDIA — GSI

THE INDIAN SPACE RESEARCH ORGANIZATION — ISRO

THE UNITED NATIONS — UN

CONTENTS

PREFACE

As editors of this volume, we wish to take this opportunity to thank COSPAR, the Indian National Organizing Committee of COSPAR, the Indian Space Research Organization, the Geological Survey of India and, most of all, the participants of the Workshop on Remote Sensing and Mineral Exploration for making the workshop a success. We also wish to thank UNESCO, COSPAR and the U.S. National Committee of the International Geological Correlation Program (IGCP) for financial support which enabled many of us to attend the meeting. The workshop was held in the Hotel Ashoka, Bangalore, India, on June 6 and 7, 1979, with approximately 100 geologists from around the world in attendance. Most of the papers presented in the workshop are contained in this volume and reflect the international "state of the art" of remote sensing in the field of geology and exploration for mineral and energy resources.

Perhaps one of the most important lessons learned during this workshop and from contacts with scientists from various parts of the world is that the availability of data ranging from simple black and white low-level aerial photography, to imagery from aerial side looking radar and thermal infrared scanners as well as satellite data is often hampered by artificial barriers based on internal national policies. There are still many countries which classify all aerial photography of their national territory, for example, as secret or confidential without considering how this restriction in use may impede national mapping and resource inventory programs. While the world-wide availability of Landsat data has done much to encourage reconsideration of such artificial barriers in many countries we urge our colleagues to press forward diligently on projects demonstrating the value of these techniques. Perhaps this publication will, in some way, help make aerial photography and other forms of data more accessible for research and resource exploration. Dr. Krishnaswamy, Director General of the Geological Survey of India, Calcutta, has stated that he considers "Landsat to be the greatest thing that has happened to mankind since the advent of Christianity." When one considers this profound remark in the light of recent history, it was truly unusual that the United States Government decided to make Landsat data available at cost to all that wish to use it. For the first time, it has enabled men to consider local problems and resource challenges in a regional context, or regions in a national context, and even national resources in an international context. As more data are collected and more cloudfree, national mosaics are completed, it will be possible to study the structure of the world, for example, in a global context. Hopefully, this will provide a better understanding of global mineral and energy distribution and lead to better exploration techniques.

We hope that the scientific results contained in this volume will contribute toward such progress by stimulating further research and cooperation throughout the world.

W.D. Carter
L.C. Rowan
J.F. Huntington

CONTRIBUTION OF LANDSAT DATA TO THE OBJECTIVES OF THE GEOLOGICAL SURVEY OF INDIA

V. S. Krishnaswamy

Director General, Geological Survey of India

I. INTRODUCTION

1.1. I feel privileged indeed to have been requested by the organisers to inaugurate this workshop on Remote Sensing for Mineral Exploration, an endeavour sponsored jointly by the COSPAR, IUGS/UNESCO, ISRO and the Geological Survey of India (G.S.I.). We, in the GSI, are already working closely with the IUGS on the IGCP Project No. 143, which is intimately related to this workshop and, also with the ISRO, on the future, National Satellite - launching programmes, in so far as provisions for earth resource surveys are concerned. We have had very fruitful contacts earlier this year with Mr. Carter, the Chairman of the Program Committee for his workshop, in connection with a Symposium on Resources Engineering, held in Bombay and I am indeed very happy that through the GSI's active contribution to this Workshop, being inaugurated this forenoon, we will be forging further links in the chain of friendship which we have already established with fellow-scientists, devoted to this newly emerging field of application of remote sensing for mineral exploration.

1.2. The Geological Survey of India, now crossing an age of six score and eight, is the second oldest scientific department in the country and third oldest Geological Survey organisation in the World. Over the past century and a quarter, this Department, which started with a strength of a single geologist, has now grown into the second largest Geological Survey organisation in the World, employing well over 2,000 earth scientists. While in the formative years of the Survey, its scope and objectives were limited largely to prospecting for minerals for the then East India Company, these have been progressively enlarged to cover many fundamental and several other applied objectives, particularly since the attainment of Independence; so much so, the GSI now is the leading agency for all types of earth science studies and research in the country.

1.3. The functions and responsibilities of the GSI have been formally enunciated by the Government of India in a charter gazetted in 1973. According to this charter, the first four objectives govern this Department's role in promoting applications of remote sensing and space research technology and these are : (1) preparation of geological, geophysical and geochemical maps of the whole country and off-shore areas; (2) exploration and assessment of mineral resources, (3) conducting studies pertaining to environmental geology and geotechnical surveys for assisting developmental projects; and (4) undertaking systematic studies and research in all sub-disciplines of earth sciences and methods and techniques of exploration and sensing.

1.4. It is now well established that remote sensing, both from orbital and aerial platforms, opens new vistas in technological development that can be fruitfully utilised for various purposes. In this gathering today, we are more concerned with its applications in surveys of natural resources, the main goal of such surveys being to gather information on the aerial distribution of different types of resources and, where possible, quantify these in terms of their availability in different periods of time, the spatial and temporal aspects of such evaluation being of cardinal importance.

1.5. It is also conceded now that the greatest advantage of remote sensing technology lies in the opportunity it affords for obtaining synoptic views of large segments of terrain and repetitive coverage at reasonable cost, thus facilitating monitoring of dynamic variations in the environment and leading to an altogether new concept of dynamic surveying. We have come to recognise, however that for optimum geo-scientific applications, satellite-derived data, being highly synoptic, is ideal mostly for coarse resolution studies of very large regions and, has to be complemented, wherever possible, with information gathered at successively lower levels through aerial photographic, aeromagnetic and ground surveys.

2. APPLICATION OF REMOTE-SENSING DATA FOR GSI WORK PROGRAMMES

2.0.1. It will now be relevant to deal briefly with the activities of the GSI in the field of applications of satellite-generated data; examine the constraints to a wider application - endeavour and analyse the short and long term objectives of our future programmes in this field. The application of satellite-generated data to the work of the GSI can be considered under 6 heads, viz. (i) structural and tectonic studies; (ii) geoseismological studies; (iii) geohydrological studies; (iv) geomorphological studies; (v) glaciological studies; and (vi) mineral resources location/evaluation studies. The order in which these are listed also seems to me to represent the order of confidence we have had in the success of the applications.

2.1. Application for structural/tectonic studies :

2.1.1. With the availability of photographs from the Gemini and Apollo missions and other satellite pictures in the late sixties of this Century, initial attempts at structural interpretation of the imagery of parts of the country, using visual analysis, have been carried out from the early 70's. The areas chosen for such analytical efforts ranged from the Himalayan border regions to the heart of the Peninsular India. From 1973 onwards, the Space Applications Centre of the ISRO started receiving some of the LANDSAT-I imageries of India and, the GSI was one of the first organisations to take up the studies of LANDSAT-I imageries. Recognising the increasing use of space imagery for geological studies in the GSI, the Photogeology Division at the six Regional Centres and at the Central Headquarters have recently been redesignated as Photogeology and Remote Sensing Division. The band of nearly 200 photogeologists available in this total set up represent the G.S.I. expertise-potential for developing Remote Sensing applications.

2.1.2. Initially, the emphasis in utilisation of satellite-generated data was in delineating major lineament patterns of the imagery as obtained on 1:1 million and 1:0.25 million scales and, in understanding their structural and tectonic significance through correlations with the mapped geological data, wherever this was possible. Subsequently, attempts were made at evaluating the geological information content on each of the different bands of the imagery and with their combinations. In some cases, with the help of the facilities available at ISRO, colour composites were also used for geological interpretation. The megalineament pattern recognised in Western India from Bombay to Delhi; the delineation of the dyke pattern in Southern Orissa and south-eastern Andhra Pradesh; and the correlation of mega-lineaments with known tectonic lines in the Jungel area of Central India, represent examples of significant results obtained from the studies of lineaments as deciphered from space imageries. The elucidation of the lineament patterns on the satellite imagery for the entire country has now been made possible through the efforts of the several Regions of the Survey and, this has resulted in the Lineament Map of India, the first draft of which is exhibited during this workshop.

2.1.3. In this lineament map, a 3-fold classification of the linears, primarily based on their length viz: mega, > 300 kms; intermediate, 100-300 kms, and minor, < 100 kms., has been adopted. It will be seen from a study of this lineament map, that while in the northern part of the country the major direction of the lineaments is ENE-WSW to NE-SW, which is at divergence with the Himalayan orographic trend as well as the principal tectonic trend, the more prominent trend in the Aravalli regions of Rajasthan and Gujarat is NNE-SSW, which is parallel to the trend of the range as well as the principal tectonic trend.

2.1.4. Stretching almost right across the sub-continent is the earlier known Narmada-Son lineament, with a characteristic ENE-WSW trend, which is spectacularly displayed in the space imageries. Extension of this lineament zone to the southern part of the Saurashtra peninsula and even across the sea has been postulated. Detailed lineament studies and their tectonic correlations in the Narmada-Son region have now been included under an inter-disciplinary project involving crust-mantle studies. This project, called 'CRUMANSONATA' initiated during this field season, 1978-79, can perhaps be described as a major beneficiary of satellite remote sensing technology and we are looking forward to the gathering of useful fundamental data in the course of execution of this project during the next five years, as a joint venture of four Regions and three Specialist Divisions of GSI at the Central Headquarters, four Indian Universities and also U.S. collaboration under the National Science Foundation.

2.1.5 In the Peninsular India south of the Narmada-Son region, the prominent lineament directions are NE-SW and NW-SE. One of the NW-SE lineaments in the southern tip of the peninsula continues on to the island of Sri Lanka.

In general, the Archaean terrain of Peninsular India shows an appreciably higher density of lineaments in comparison to the fold-belts of the Himalayas which, in turn have a much higher lineament density as compared to the alluvial tract of the Sindhu-Ganga-Brahmaputra Basins.

2.1.6. We are not in a position yet to claim that all the lineaments we have patiently deciphered from space imagery have been satisfactorily correlated with known or reasonably postulatable geological or tectonic features. In fact, in many cases, my own personal experience has been that there are puzzling arrays of lineaments in several directions, for which no logical or fact-supported explanations are forthcoming. In a few cases, however, our attempts have been rewarding. Thus, the structural and tectonic correlations attempted on the basis of style and trends of lineaments as well as their density distribution collated with the ground truth data collected during the course of geological surveys over the years, have led to the elucidation of four major tectonic events in the Eastern Ghat belt. In a second case history dealing with the Godavari, Pranhita valley, concurrent synthesis of field data with satellite imagery and airborne data has led to the concept of a two stage evolution of the graben containing extensive coal horizons. A third case-history of conjunctive interpretation of satellite and ground-data is the one relating to Karnataka which will be presented in greater detail during this workshop.

2.2. <u>Application for geo-seismological studies</u>

2.2.1. The lineaments as interpreted have been matched with the locations of epicentres of past earthquakes, in studies carried out by the Northern and North-eastern Regions of the Survey. Except where vertical tectonic features are involved in the focal mechanism, such two-dimensional correlation of epicentral locations with lineaments could be misleading, particularly with low-angled Himalayan thrusts. In such cases, focal depth data have also to be taken into account, before effecting correlations with specific lineaments.

2.2.2. Pending such detailed evaluations, it is tentatively concluded that the epicentres in the Western Himalaya are mostly correlatable with the NW-SE or WNW-ESE trending lineaments. Likewise, in the North-eastern Himalaya, it is tentatively interpreted that some of the Landsat-lineaments lie in zones of high seismic activity.

2.3. <u>Application for geohydrological studies:</u>

2.3.1. In the Anantapur district of Andhra Pradesh, the lineaments picked up from aerial photographs and Landsat imagery were located on the ground and tested by seismic refraction methods to assess their suitability for ground-water exploration. Subsequent exploration by drilling at the more favourable locations encompassed by the lineaments gave encouraging yields of ground-water.

2.3.2. Similarly, in Coimbatore district of Tamil Nadu and adjoining Palghat district of Kerala, in areas traversed by major lineaments, identical pattern of studies undertaken in collaboration with the Central Ground-water Board have lead to the location of well-sites with spectacular yields of ground-water in this hard rock area.

2.3.3. Another aspect of application of Landsat-data generated lineament map is the correlation with hot spring locations with the lineaments thereby, aiding regional geological assessment of geothermal potential. This endeavour has been initiated in the Southern, Central and Northern Regions of the Survey.

2.4. Application for geomorphological studies

2.4.1. In the studies carried out by the Northern Region, and with visual interpretation in four spectral bands, the geomorphological evolution of the Ghagra and Gomti river basins of Uttar Pradesh and the Sahibi river basin of Haryana has been clearly brought out. This information has been of considerable utility in the Quaternary geological mapping programmes of the GSI.

2.4.2. A very unusual application of the Landsat-I imagery interpretation was the deciphering of the old confluence point of the Ganga and Yamuna rivers near Allahabad, U.P., which, in Hindu scriptures, has been designated as the "Triveni Sangam" (or tri-river confluence). It is considered possible that this "Sangam" was earlier, 25 km. away to the south of the present "Sangam", as adjudged from the evidence of channel migration of the Ganga from south to north since the Vedic times.

2.4.2. Geomorphological maps have also been prepared with the help of Landsat data for environmental development studies of Chandrapur district, Maharashtra and Anantapur district, Andhra Pradesh and for regional geological studies of southern Rajasthan.

2.5 Application for Glaciological Studies

2.5.1. Visual interpretation of Landsat-Imagery has been successfully applied to the study of snow-cover in the upper reaches of the Beas river basin, Himachal Pradesh as a supporting programme for the glaciological studies undertaken by the Survey. An examination of the enlargements of the imagery taken in March, 1972 enabled the differentiation of various categories of snow accummulation, such as, wet snow and permanent snow and, calculation of areas covered by these categories. Ground stations have been proposed for correlation between snow-cover and water discharge, as well as with other parameters. Repetitive imagery studies and correlation of data with ground truth information is expected to lead ultimately to a snow melt run-off prediction model for effective planning and management of water resources available from the Himalayan glaciers.

2.6. Application for mineral resource location evaluation studies.

2.6.1. The Geological Survey, through its various Regional Centres for photogeology and remote sensing, is vigorously participating in the IGCP Project No. 143 dealing with remote sensing and mineral exploration and the results of some of these endeavours will come up for discussion in this workshop. Although many lineaments or intersection of lineaments, as deciphered from space imageries, have been correlated with known centres of mineralisation under the Indian set-up, in almost all these cases, the arguments advanced are "a posteriori". To the best of my knowledge, we have not been able to demonstrate an " a priori" case that has lead us to the location of a mineral deposit as a result of study of the lineament patterns. On the other hand, in the case of one Project for mineral search, initiated by me on the basis of space imagery/airphotograph interpretations of a striking lineament - the Asmara lineament - running for more than 100 kms. from Allahabad to Panna in Central India, subsequent geophysical work has not given any clues of significant mineralisation available at shallow depths of 150 m. or so. This does not, however, rule out deeper causative factors for the lineament or deeper locations of mineral deposits, if any.

2.6.2 However, one fairly successful interpretation can be cited. In applying the imagery data for mineral resource location, the case of the diatremes of southern Andhra Pradesh and the Panna-Jungel area of Central India deserves mention. These diatremes have been interpreted to be located at or close to the intersections of prominent lineaments. With this as a starting point, geomorphological studies are now being extended to delineate potential diamond-bearing gravels and placers along palaeo-drainage channels. Satellite-data and aerial photography taken together, seem to offer vital clues in such studies.

2.6.3. Although the known base metal deposits in the Himalayas are found to be in proximity of intersections of lineaments, the efficacy of this as a guide to mineralization requires further study and, this would perhaps be one of the topics discussed later during this Workshop. The presence of a circular structure in the Archaean gneisses of Mikir hills, Assam, noted on Landsat imagery has led to the location of ultramafic rocks by subsequent ground survey, and further studies in this regard are under way.

3. PROGRAMMES FOR THE DEVELOPMENT OF SPACE RESEARCH APPLICATIONS IN G.S.I.

3.1. With the modest achievements made in India so far and, with the background of experience we have gained in the applications of orbital imagery data for GSI's objectives, during the last 6 years or so, it is possible to formulate now, realistically, the constraints and expectations in regard to the future programmes of the Geological Survey from Landsat and other developments in satellite technology.

3.2. In the three main functional areas of satellite remote sensing viz., data acquisition, formating (including processing), analysis and interpretation, the Regional Centres for Photogeology, and Remote Sensing require strenghthening in regard to the first two areas, so that improved facilities are available to our investigators in this field of specialisation.

3.3. Though a number of computer-based systems for analysis and interpretation for resources inventories have been designed and marketed, these are fairly successful mostly in applications involving landuse, vegetation and forest mapping. In the case of geology, and to a large extent soils, the analysis being much more subjective and ground data-controlled, most of the work so far carried out has mainly been in the field of visual interpretation and analysis of photographs or photo-format imagery. In view of the large variations in the spectral response and, consequently, of spectral signatures of geological targets, many of the automated data handling techniques, so far experimented upon, have to be restricted to the different type of pre-processing and enhancement, to yield more easily interpretable imagery.

3.4. In this context, if tt is possible to design and fabricate automated systems, at least partially indigenously, these could be utilised for the preparation of ratio images, principal component transforms, edge enhancements, enumeration of areas, and similar endeavours in the case of satellite multi-spectral data and, for correcting for planimetric distortions conjointly with airborne multi-spectral data. Concurrently with this, a programme of collection of spectral response data in the field and in the laboratory, the former from very low altitudes, under varying atmospheric conditions and on the diverse rock and terrain types in the country, has to be launched so that meaningful correlations could be established with temporal imagery. Such a programme of data collection would require compatible instrumentation, trained personnel, access to target areas and adequate data processing facilities, spread all over the country.

3.5. The GSI will be taking action on fulfilling the aforementioned requirements for more effective space applications research endeavour, by launching a) programme of acquisition of equipment for formating in the Regional centres; b) training of some of our scientific personnel in the required technology and c) procurement spectroradiometers and other instruments from abroad. Simultaneously, the Geological Survey has also taken up the initiative of collaborating with the ISRO and other organisations in the field of electronics instrumentation, for indigenous development of field and laboratory model instruments in adequate numbers. Our aim is to deploy 30 to 40 field parties from about 6 to 8 Regional Centres, for collection of ground spectral response data. The Regional Centres will also be equipped with adequate laboratory instrumentation and data processing facilities. It is expected that in the near future the Department will be able to develop a sufficiently large data bank of spectral signatures, that could be utilised in full measure for further development of space-data applications in the field of geoscientific endeavours in the country.

3.6. The experience gained so far with imagery from Landsat-I indicates that preliminary reconnaissance level maps of areas of difficult access could be made quickly and at reasonable cost savings. Also interesting is the fact that most of the known ore deposits are located in the vicinity of intersections of Landsat-identified lineaments and, some deposits, whose association with known structural controls not evident earlier, also appear to be located close to these features. If only we are able in future, to use the above "a posteriori" deductions to lead us to "a priori" discoveries of mineral resources, we would have taken a great leap forward in furthering one of the charter objectives of the GSI.

3.7. A spin-off of application of Landsat data on the scale of 1:1 million is the fillip that it has given to broader conceptual formulations on the relationship of mineral deposits to the regional geological framework. This widening of the conceptual level may itself help to carve new paths amidst the jungle of conventional thinking to new ideas that may lead us to new discoveries. Should this anticipation materialise, this would be no mean achievement for the space technology application.

3.8. For the planner-in-charge of mineral resource inventorying, the scale range from 1:1 million to 1:0.25 million is optimal for modelling and assessment of the resources picture. Therefore, it is no wonder that the most comprehensive commercial use of Landsat data, so far, has been in exploration geology, by the mining and petroleum community. The geologist's background and training in projecting surface geological data for understanding the underlying controls of mineralisation is one of the principal reasons for such a lead in space applications to exploration geology. We can only hope that this lead is maintained through the postulation of new ideas based on new concepts generated by satellite data.

3.9. If I could take the liberty of projecting the possible future applications of Landsat and its successor programmes, in regard to geoscientific activities in this country, these may fulfil, a majority, if not all, of the following :

1) Development of geological maps of areas not previously covered and refinement of existing maps as to their accuracy and completeness in several details;

2) Identification of large-and medium-scale geological and structural features and their correlation over widely separated areas;

3) Provision of preliminary geological appraisal for planning and location of transportation, communication, irrigation, energy and industrial projects;

4) Selection of potential mineral bearing areas for detailed examination by airborne geophysical and ground studies;

5) Provision of greater opportunities to monitor, through use of repetitive coverage, transient geological features such as changes in river courses, glacial and desertification features and, their impact on developmental activities. The main attraction in use of satellite derived data for all the objectives as listed above, will naturally, be, the possibility of considerable saving in the time required for these activities.

3.10. The Geological Survey of India is convinced that remote sensing can and will play a crucial role in the area of monitoring developmental activities dispersed widely in space and time. To ensure this, an integrated system comprising of satellite based surveys at close frequency, airborne surveillance of selected smaller target areas and ground data collection in key regions would be optimal, augmented by rapid data transmission, where necessary, to the processing centres, through satellite relays. During the coming decade we hope to reach the stage when the compilation of geological maps of any area, whether of small scale or large scale, would have involved, at some stage or other, the use of satellite and airborne imagery as well as aerial geophysical data, and, in many cases, with correlations based on spectral response patterns.

3.11. While the synoptic views, obtained through satellite remote sensing, will have considerable impact on regional geological mapping, detailed remote sensing missions are also expected to prove useful in providing the linkages to anomalous patterns, which alone are indicative of mineral occurrences. With adequate recognition of the benefits of and the limits as well as the constraints to the utilisation of remotely sensed data and, with the supplementation of these data in an ever increasing measure and on larger scales, with information derived from other, aerial and ground investigatons, there is no doubt that this scenario of rapidly emerging technology would offer expanding opportunities for the earth scientists in fulfilling their accredited roles.

3.12. The Geological Survey of India, as the premier organisation of geoscientists in this country would, therefore, like to go ahead with its future programme of building up its capabilities in the field of remote sensing applications, to a measure and to a scale compatible with our vast programme of ground geological mapping, with which 50% of the country still remains to be covered on 1:50,000 scale and to subserve our objective of regional mineral assessment and other resource developmental activities. We offer to share our expanding experience in this field with the scientific community spread out all over the world, realising that only by such joint analysis in depth, both of our achievements and failures in this field, can the know-how in this emerging field of space technology, be improved upon and made more beneficial for our future endeavours.

3.13. From this point of view, all the geoscientists of Geological Survey of India deeply appreciate the fact that this important International Workshop, organised by the COSPAR and INCOSPAR is being held in India. We look forward to benefiting by the formal and informal contacts developed with the fellow-scientists assembled here, both in the course of this Workshop, and thereafter, thereby, taking forward, by many steps, the fulfilment of our National aspirations and objectives, for which this 128 year old National Resource Survey Organisation has been in existence. I have very great pleasure, therefore, in inaugurating this Workshop on Remote Sensing for Mineral Exploration.

OBJECTIVES OF THE WORKSHOP

W. D. Carter

U.S. Geological Survey, Reston, Virginia 22092, U.S.A.

Gentlemen: I wish to take this opportunity to thank you all for your interest in attending this workshop on remote sensing and mineral exploration. It is part of the International Geological Correlation Program of the International Union of Geological Sciences, a UNESCO/ICSU organization. It has been my pleasure to serve as the IUGS representative in COSPAR for the past 6 years and it is partly from this participation that the IGCP Project 145 "Remote Sensing and Mineral Exploration" developed about three years ago. While this IGCP project has participated in many other scientific meetings throughout the world this is our first dedicated workshop. For several reasons it is appropriate that it be held in India. Geologists of the Geological Survey of India have been active in our IGCP Project since its beginning. India is about to open its Landsat reception station to take better advantage of the uses of satellite data in resource mapping, inventory and development. It is therefore most appropriate that we meet here to discuss what we have accomplished and where we are going in exploration technology. Our purpose is to develop and leave a model for others to follow. I wish to thank COSPAR and the Indian Organizing Committee for agreeing to provide the time in its very busy program for this opportunity to bring you all together. Dr. P.D. Bhavsar and Dr. D.S. Kamat of ISRO and Dr. J.G. Krishnamurthy of the Geological Survey of India have been most helpful in arranging local affairs, materials and space for us. They deserve our warmest thanks. UNESCO and COSPAR generously provided funds to help defray some of the costs of travel for some of our visiting scientists whom you will hear. The U.S. National Science Foundation and National Academy of Science have also contributed to this effort as have several other national agencies represented by our participants from other countries.

During these two days we have an opportunity to share our experiences in the use of satellite and other remote sensing techniques in geologic mapping and mineral and energy exploration. I will consider the workshop a success if we can accomplish the following:

1. Use this opportunity to get to know each other on a personal basis to strengthen the bonds of our profession and establish better scientific exchange.

2. We will learn more about our progress, our problems, and our plans for the future. By this method, we may, in some way, help each other to move faster down the complex and difficult road of finding resources that can help meet the growing needs of mankind.

3. The publication of our findings will serve as a benchmark report on where we are at this time in the history of development of satellite technology. Hopefully, those who follow us will (a) avoid some of the pitfalls we have discovered and (b) proceed toward optimum use and application of this technology.

4. Let us use this opportunity to think of and, perhaps, define new experiments to test the available systems on mineral commodities that are critical to Man's future needs.

Finally, I wish to close this introduction by putting our efforts to develop the use of aerospace technology into proper perspective.

1. Satellite images are providing new views of the earth that can make the geologist's job of finding mineral and energy resources more efficient. They will not, however, replace the needs for traditional field mapping or geophysical and geochemical survey methods.

2. Experimental methods of merging Landsat data with other survey techniques such as aeromagnetic data should be encouraged to develop until they become operational tools for exploration.

3. Visual interpretation of Landsat images has proven highly useful in geologic applications. The uses of repetitive coverage and digital analysis methods are rapidly developing. We should also be thinking of how to design ways of integrating digital Landsat data with other digital information and integrating them into computer-based mineral and energy resources information systems.

PRINCIPLES OF ELECTROMAGNETIC RADIATION AND REMOTE SENSING

L. C. Rowan

U.S. Geological Survey, Reston, Virginia, U.S.A.

Dr. Rowan reviewed the principles of geologic remote sensing with emphasis being placed on lithologic mapping. An outline of his illustrated lecture is given below along with a list of published references which are of current use.

LITHOLOGIC MAPPING

I. Properties

- A. Gamma radiation
- B. Fluorescence
- C. Spectral Reflectance (0.4-2.5 um)
  1. Visible and short wavelength Near-IR (.4-1.3 um) -electronic transitions in Fe, Mn, Zn, etc.
  2. Long wavelength Near-IR (1.3-2.5 um) -vibrational overtones in water, hydroxyl-bearing minerals and carbonates.
- D. Spectral Emittance (8-14 um)
  1. Fundamental vibrations in silicates.
- E. Thermal Inertia (8-14 um)
- F. Vegetation
  1. Plant geography
  2. Spectral features
- G. Surface Roughness

II. Applications

- A. Spectral Reflectance (.4-1.3 um)

1. Fe transitions and absorption bands
   -limonite, hematite, geothite spectra
   -origin of visible color
   -dominance in Landsat MSS response region
   -altered/unaltered rock spectra

2. Processing techniques for displaying Fe related spectral relectance differences
   -use of ratio images for subduing reflectance variations due to albedo and topography.
   -construction of color-ratio composite images to show spectral reflectance differences in color.
   -other techniques including classification and principal component analysis.

3. Use of color-ratio composite image for mapping
   -extraction of colors using visual, color-recognition scanner, and Munsell algorithm.

4. Limiting factors
   -vegetation; total amount and affects on ratio and contrast stretch selection.
   -topographic affects.
   -atmospheric considerations.
   -mineralogy of altered rocks
   .non-limonitic altered rocks (Cuprite, Nev.)
   .limonitic unaltered rocks (Coaldale, Nev.)

B. Spectral Reflectance (1.3-2.5 um)

1. Location and origin of absorption bands
   -exclusion of 1.4 and 1.8 um bands due to atmosphere.
   -2.2 um band related to clays, alunite, muscovite and other hydroxyl-bearing minerals.
   -2.35 um carbonate band and its infrequent appearance.

2. Use of 2.2 um band for mapping altered rocks
   -NASA 24-channel scanner images.
   -Cuprite, Nev.; nonlimonitic altered rocks.
   -Coaldale, Nev.; limonitic unaltered rocks.
   -East Tintic Mt.; detection of altered rocks; no confusion with carbonates.

C. Spectral Emmittance (8-14 um)

1. Location, origin and variation of silicate absorption feature in lab spectra; lack of absorption feature in carbonate spectra in this region.

2. East Tintic Mt. color composite
   -conspicuous quartzite
   -inconspicuous limestone
   -differentiation of silicified and argillic altered rocks.
   -importance of vegetation.

3. Possible limitations.

D. Thermal Inertia (8-14 um)

1. Definition, diurnal cycle, annual cycle, need for a model.
2. Mill Creek, OK and Oman cases.
3. Limitations - vegetations, soil moisture.

III. Future Prospects

A. High-spectral resolution
B. Stereoscopic multispectral images
C. Active systems
D. Fraunhofer-line discriminator (FLD).

REFERENCES

Reeves, R.G., Ed., 1975, Manual of Remote Sensing: American Society of Photogrammetry, Falls Church, Virginia, U.S.A., Vol. I, 867p., Vol. II, 2144p.

Sabins, Jr., F.F., 1978, Remote Sensing Principles and Interpretation: W.H. Freeman and Co., San Francisco, Calif., U.S.A., 425p.

Smith, W.L., ed., 1977, Remote-Sensing Applications for Mineral Exploration: Dowden, Hutchinson and Ross, Inc., Stroudsburg, Pa., 18360, U.S.A., 391p.

CHARACTERISTICS OF THE LANDSAT SYSTEM AND DATA FOR GEOLOGIC APPLICATIONS—AVAILABILITY OF DATA

W. D. Carter

U.S. Geological Survey, Reston, Virginia, U.S.A.

SUMMARY

The experimental earth resources technology satellite series, formerly called ERTS and now known as Landsat -1, -2, and -3 were launched into polar orbit by the United States National Aeronautics and Space Administration (NASA) in July 1972, January 1975, and March 1978, respectively. Landsat-1 operated successfully until January 1978 and provided thousands of images of the land surface that have been studied by hundreds of scientists throughout the world. Landsat-2 is currently operating but has lost its capability to record data from areas outside of the range of existing ground stations. Landsat-3 is also operating well except for its thermal infrared band (10.4 - 12.6 m) which failed a few months after launch. Details on the characteristics of the satellites and their data, summarized in Table 1, are contained in a wide variety of publications among which are the "Landsat Data Users Handbook," available from the EROS Data Center, Sioux Falls, South Dakota, USA, "ERTS-1: A New Window on Our Planet," edited by Williams and Carter (1976) and "Characteristics of the Landsat Multispectral Data System" (Taranik, 1978) distributed by the U.S. Geological Survey, Reston, Virginia, USA.

Although the Landsat series are still considered to be experimental satellites their data are being used in a variety of operational programs throughout the world. The need for such data is exemplified by the investment of national agencies of Canada, Brazil, Argentina, Italy, Sweden, Iran, Upper Volta, India, Australia, and Japan, which have built or are building Landsat reception stations to ensure continuity of data over their territories and neighboring regions. In order to construct such stations these countries sign an agreement with the United States to sell their data at a reasonable cost to all who wish to use it. Although the United States has not yet committed itself to an operational Landsat program international response and interest in the Landsat program has strongly endorsed the value of the experimental program and have greatly aided in determining its future. Several government-wide studies are now underway which are considering several alternative ways of conducting an operational program. Several bills proposing an operational program are being considered by the United States Congress and it now appears that some action will be taken during the current year.

In the meantime Landsat-D, the fourth of the Landsat series, is being developed for Launch in 1981 from a Space Shuttle sortie mission. Landsat-D is designed to be a more advanced, higher resolution system which includes a 5-band multispectral scanner system (MSS) similar to that on Landsat-3 and a 7-band thematic mapper (TM). The characteristics of Landsat-D are summarized in Table 2. Significant changes include the addition of a solar infrared band in the 2.08 - 2.35 m range which will enable geologists to map areas of clay minerals having hydroxyl ions. The orbital characteristics of Landsat-D are different from the others in that it will be at a lower altitude (705-720 km) and collect data in a different sequence. The swath width of the images, however, will continue to be 185 km to maintain continuity with data already collected. The higher data rate (80 mbps) designed to provide better resolution (30 m) of the thematic mapper (TM), however, represents a significant increase over previous Landsats and will require modifications at existing reception stations if these data are to be received.

Worldwide coverage of Landsat data is available at a modest cost to all who wish to purchase it through the EROS Data Center, Sioux Falls, South Dakota. Those nations having receiving stations also distribute data at cost.

Because Landsat data provides synoptic, orthographic views of the earth on a worldwide basis geologic applications of Landsat data have developed on a broad base. In the area of mineral and energy resource exploration they range from local to regional and continental studies of lineaments relating to tectonic disturbances and mineral, oil and gas distribution using single Landsat images and mosaics. Some have focussed on the relationship of structural features to mineral belts and metallogenetic provinces. Others deal with the structures of sedimentary basins and platforms and the distribution of petroleum and gas deposits.

Digital analysis of Landsat data deal with the spectral discrimination of rocks, gravels, and soils; the separation of altered and unaltered rocks; and the relationship of vegetation poisoned by underlying mineralized zones. Examples of all of these applications are found in the "Bibliography of Remote Sensing and Mineral Exploration that is being developed under IGCP Project 143 and now contains over 700 citations relating to the project.

The wealth of remote sensing literature and developing global expertise in the geologic uses of satellite data now make it possible to conduct comparative studies of similar rock types (e.g. Precambrian system) on a global basis. There is no reason why these same techniques cannot be applied to certain types of mineral deposits, especially to those deposits which tend to occupy large surface areas (e.g. iron, coal, phosphate).

Table 1. Characteristics of Landsat 1, 2, 3
Initially called Earth Resources Technology Satellite (ERTS)

LAUNCH DATES:	Landsat-1, July 23, 1972. Operation ended January 6, 1978 Landsat-2, January 22, 1975 Landsat-3, March 5, 1978
ORBITAL ELEMENTS:	Orbit: Circular, near polar Inclination: 99.09 Altitude: 919 km Coverage: 82 north to 82 south Period: 103 minutes, crossing Equator at 0930 hours, local time Cycle: 18 days. Landsat-3 follows Landsat-2 by 9 days.

SENSORS:

	Wavelength (μm)	Resolution	Image format
Return Beam Vidicon Cameras (RBV)			
Landsat-1 and -2, three RBVs:			Simultaneous view from 3 cameras of scene
Band 1	0.475-0.575 (blue-green)		185 x 185 km
Band 2	0.580-0.680 (yellow-red)	80 m	14% sidelap at Equator
Band 3	0.690-0.830 (red-IR)		10% forward lap
Landsat-3, two RBVs	0.505-0.750 (panchromatic)	40 m	2 side-by-side images 98 x 98 km (4 RBV images coincide nearly with one MSS frame)
Multispectral Scanner			
Landsat-1, -2, -3			
Band 4	0.50-0.60 (green)		185-km strip image framed with 10% forward lap
Band 5	0.60-0.70 (red)		
Band 6	0.70-0.80 (near	80 m	14% sidelap at Equator,
Band 7	0.80-1.1 infrared)		increasing toward poles
Landsat-3 only:			
Band 8	10.4-12.6 (thermal IR)	240 m Range: 260°K-340°K	

2 wideband tape recorders (only one tape recorder operating on Landsat-2)

Data Collection System (from up to 1000 platforms, each with 8 sensors) Terminated on Landsat-2 on launch of Landsat-3. Reinstated on Landsat-3.

DATA AVAILABILITY: All data received in the U.S. are in the public domain and can be purchased either as image products or as computer-compatible tapes from the Department of the Interior's EROS Data Center, Sioux Falls, South Dakota.

The EROS Digital Image Processing System (EDIPS) became operational in late 1978. High density digital tapes supplied by NASA/GSFC are converted by EDIPS high resolution laser-beam film recorder to first-generation products in 241-mm format, enabling EDC to deliver second-generation products to data users. CCT data from HDTs are formatted and recorded by EDIPS.

GROUND DATA ACQUISITION STATIONS: Existing: Fairbanks, Alaska, Goldstone, Calif., Greenbelt, Maryland, in the U.S. Prince Albert, Saskatchewan, and Shoe Cove, Newfoundland, in Canada; Cuiaba, Brazil; Fucino, Italy; Shahdasht, Iran; Marchicita, Argentina; Alice Springs, Australia; Hyderabad, India; Tokyo, Japan; Kiruna, Sweden; Ouagadougo, Upper Volta; Agreed upon but not yet operating: Kinshasa, Zaire.

FOR MORE INFORMATION:

Landsat Data Users Handbook
Missions Utilization Office
Code 902, NASA/GSFC
Greenbelt, Md. 20771

Landsat Data Users Notes
Users Services
EROS Data Center
Sioux Falls, S. Dak. 57198

Landsat-3 Reference Manual
General Electric Space Div.
P. O. Box 855
Philadelphia, Pa. 19101

Table 2. Characteristics of Landsat-D

EXPECTED LAUNCH DATE: Early 1981, in Space Shuttle

ORBIT: Circular, near-polar
Altitude: Not yet determined (705-720 km)

SPACECRAFT: Multiple-Mission Modular Spacecraft (MMS)
Greater attitude stability (to within 0.01° compared to 0.7° for earlier Landsats) will improve geometric accuracy of data

PROPOSED SENSORS (with improved spatial, spectral, and radiometric resolution):

Multispectral Scanner (MSS)	Wavelength(μm)	Resolution	Swath width	Scenes/day	Data rate
5 bands					
green	0.50-0.60				
red	0.60-0.70			200 (U.S.)	
solar IR	0.70-0.80	80 m	185 km	400 (World-wide)	15 mbps
solar IR	0.80-1.1				
thermal IR	10.4-12.6	240 m			
For continuity of Landsat data					
Thematic Mapper (TM)					
7 bands					
blue-green	0.45-0.52				
green	0.52-0.60			50 (U.S.)	
red	0.63-0.69	30 m	185 km	100 (World-wide)	84 mbps*
solar IR	0.76-0.90				
solar IR	1.55-1.75				
solar IR	2.08-2.35				
thermal IR	10.4-12.5	120 m			

*Large volume of data from TM may require use of image compression techniques.

GROUND DATA SYSTEM: Will improve worldwide coverage by eliminating unreliable tape recorders and shorten time frame of delivering data to users.

Direct Readout Stations (DROS):
Existing, or with existing agreements - Prince Albert, Saskatchewan, and Shoe Cove, Newfoundland, Canada; Cuiaba, Brazil; Fucino, Italy; Shahdasht, Iran; Argentina, India, Australia, Japan, Sweden, Upper Volta.

Being negotiated - Zaire
Downlinks: MSS - S-Band (2 gHz)
TM - X-Band (8 gHz)

Tracking Data and Relay Satellite System (TDRSS): To ground station at White Sands New Mexico, to GSFC via Domsat for preprocessing, to EROS Data Center and other users.

FOR MORE INFORMATION: Resource and Environmental Surveys from Space with the Thematic Mapper in the 1980's, 1976, Committee on Remote sensing Programs for Earth Resource Surveys, National Research Council, available from NTIS as NRC/CORSPERS-76/1, 122 p.

CONTACT: Dr. V. V. Salomonson, Landsat-D Project Scientist
NASA GSFC, Code 913, Greenbelt, Maryland 20771

DESCRIPTION AND STATUS OF INDIAN LANDSAT RECEIVING STATION AND DATA AVAILABILITY

K. R. Rao

National Remote Sensing Agency, Secunderabad, India

ABSTRACT

The use of satellite remote sensing data for earth resources survey is well recognized. The National Remote Sensing Agency (NRSA), which has been established to provide resources survey services to users in India employing remote sensing techniques, has undertaken and completed several operational remote sensing surveys of natural resources using its specially equipped aircraft, sophisticated interactive computer systems, comprehensive modern photo processing laboratory, well-trained multidisciplinary teams of scientists/ engineers. Both the aerial and satellite remote sensing data were used. The Landsat data was being obtained from the EROS Data Center, USA. In order to obtain comprehensive and time effective data of natural resources of Indian sub-continent and also improve the effectiveness of NRSA in conducting operational remote sensing surveys, the Government of India authorized the NRSA to set up an Earth Station to receive earth resources data in the form of digital signals directly from the Landsat series of satellites launched by NASA, USA. In connection with this, a Memorandum of Understanding was signed in January 1978 between the Government of India and United States of America. The Earth Station, being constructed near Hyderabad, is capable of acquiring, tracking, receiving, recording, monitoring, preprocessing and producing Landsat 2 and 3 MSS data, Landsat 3 RBV data and TIROS-N AVHRR data.

This talk covers the following aspects of the NRSA Landsat Earth Station:

(a) Site survey for locating the Earth Station
(b) Infrastructures built
(c) Antenna receiving and data processing systems
(d) Data Dissemination.

A site near Hyderabad, in Andhra Pradesh was selected after carrying out a site selection survey keeping in view of various technical, civil and other constraints. The infrastructure built includes operations and control building with necessary power equipment, antenna pedestal and boresight facility. The major equipment of earth station, viz., antenna receiving system (ARS) and Data

Processing System (DPS) are being procured from two US companies. One of the significant features of the deal with these companies, is participation of NRSA engineers in design, development and manufacture of ARS and DPS equipment. As a result, our engineers, who would be knowing the systems thoroughly, will be in a position to commission, maintain and operate the system, and if necessary modify and upgrade them.

The Earth Station, which will be operational shortly, will provide all standard output products.

STATUS AND PLANS OF SEO SATELLITE AND RECEIVING STATION

D. S. Kamat

Remote Sensing Area, Space Applications Centre (ISRO), Ahmedabad 380 053, India

ABSTRACT

This paper describes the data products concerning the vidicon camera payload data , the passive microwave radiometer payload data, and the data collection platform data to be collected from the Satellite for Earth Observation (SEO) which Indian Space Research Organisation is to launch in the near future. This paper also describes, in brief, SEO satellite and ground receiving system for data reception when SEO is in orbit.

INTRODUCTION

The Satellite for Earth Observation (SEO) to be launched in June 1979 is a low orbiting satellite at an altitude of 525 Km with an inclination of 51°. The satellite will carry payloads for applications in both earth resources survey and environmental monitoring as well as certain other experimental/technological payloads such as X-ray sky monitor, heat pipe, thermal control coating package, and solar cells panel.

Two ground stations, one at SHAR and the other one at Ahmedabad will acquire the data in real time on ground command.

The SHAR ground station will have the following facilities:

1. Reception of LBT and HBT data using a steerable monopulse yagi antenna with servo control and tracking receiver and recording on computer compatible tapes.

2. Tracking of SEO by tone range, range rate and interferometry systems.

3. Data display for quick-look.

4. Real time display and control of spacecraft status.

5. Near real time attitude determination.

6. Telecommand.

The Ahmedabad ground station will have the data reception facility with a manually steerable antenna, analog magnetic tape recording, tracking by tone ranging system, quick-look display system and limited spacecraft health monitoring facility.

It is proposed to describe here the various data products of SEO and the format of each one of them.

The different applications oriented sensor systems on SEO are -

a) TV Camera System : This consists of two vidicon cameras, one operating in the visible band 0.54 - 0.66 micrometers and the other operating in the near IR band 0.75 - 0.85 micrometers.

b) Microwave Radiometer System : This is referred to as SAMIR (Satellite Microwave Radiometer) and consists of three radiometers, two of which are operating at 19.35 GHz and the third one operating at 22.235 GHz.

c) Data Collection Platform System : This consists of eight data collection platforms (DCPs) spread over the country, which transmit meteorological data at random repetition times.

The different data products presented here are :-

i) SEO TV Photographic Data Products (Browse and 70 mm formats)

ii) SEO TV CCT Data Product

iii) SEO SAMIR Data Product

iv) SEO DCP Data Product.

SEO TV PHOTOGRAPHIC DATA PRODUCTS

Two forms of photographic data products are available for the SEO TV Sensor data. First is the 'browse' data product where the data is only radiometrically corrected and the second is the 70 mm data product which is both radiometrically and geometrically corrected. In this section the information provided on these photographic data products is explained.

A sample browse data product is illustrated in Fig.1(a). Its size is such that each image can be mounted on a 35 mm slide and projected for quick-look analysis.

First Annotation Line at the Top (Left to Right)

ISRO	: Name of the Organisation
SEO	: Name of the Project
083	: Day number since launch
11085	: Coded time of exposure.
	11 Hours (First Two Digits) 08 Minutes (Next Two Digits) 5 Tens of seconds (last digit)
1	: Spectral band number. 1 for visible, 2 for near IR.

Second Annotation Line at the Top (Left to Right)

23-12-78	: Day of exposure, Date-Month-Year
1	: Beam current indication (1,2 or 3)
2	: Exposure duration. 1 for 1 ms, 2 for 1.5 ms and 3 for 2 ms.
0946	: Revolution number
139	: Satellite heading in degrees as measured from true North in the clock-wise direction.

Annotation Line at the Bottom (Left to Right)

23N37	: Latitude of the subsatellite point (23 degrees and 37 minutes)
71E38	: Longitude of the subsatellite point (71 degrees and 38 minutes)
E60	: Sun elevation, degrees
A120	: Sun azimuth, degrees

An equal 8 step grey wedge is provided at the bottom.

Next, the information provided on the 70 mm product is explained. A sample product is presented in Fig.1(b) and the various annotation characters around the imagery are explained below :-

First Annotation Line at the Top (Left to Right)

C	: Classification. C signifies classified and its absence unclassified imageries.
B1	: Processing status. B for bulk and P for precision processing. 1 indicates sub-status for future use.
10-09	: Row and column numbers of the grid cell in which the subsatellite point is located (see Fig.2)
0946	: Revolution number
S	: Satellite heading. N for North-bound and S for South-bound passes.
139	: Satellite heading in degrees as measured from true North in the clock-wise direction.
2	: Exposure duration. 1 for 1 ms, 2 for 1.5 ms and 3 for 2 ms.
A	: Image acquisition site. A for Ahmedabad and S for SHAR.
ISRO/SEO1	: Constant, indicating Organisation and the Project.
78-12-23	: Day of exposure. Year - Month - Date
11-08-53	: Time of exposure. Hour-Minute-Second
1	: Spectral band number. 1 for visible and 2 for near IR.
083	: Day number since launch.

Last Annotation Line at the Bottom (Left to Right)

S	: Constant, signifying Subsatellite Point.
N21-47	: Latitude in degrees and minutes.
E71-38	: Longitude in degrees and minutes
C	: Constant, signifying Format Centre

N21-47 : Latitude in degrees and minutes
E71-38 : Longitude in degrees and minutes.
SUN : Constant, signifying Sun Angle
EL60 : Elevation in degrees
AZ120 : Azimuth in degrees

Annotation Around the Border of the Imagery

The latitude and longitude of the corner and middle points on all the four sides of the imagery are also provided.

For example, the middle point on the right side is indicated as :

N22-47 : Latitude in degrees and minutes
E72-52 : Longitude in degrees and minutes.

Step Wedge on the Extreme Left :

A sixteen step gray wedge is provided on the extreme left of the format. Out of the 128 gray levels in which SEO data is available, the range 0-127 is divided equally into sixteen intervals to represent the grey wedge and the level 127 is represented by the white background on a positive imagery.

Identifying Image Location

For easy location of the imagery with respect to its subsatellite point on the map of India, a 40 x 40 grid has been super-imposed as shown in Fig.2. The grid extends from 0° North to 40° North in latitude and 57° East to 97° East in longitude and is comprised of one degree interval divisions. The row and column numbers, as shown in the annotation format of the photographic data product, are also indicated in Fig.2.

SEO TV CCT DATA PRODUCT

One of the data products of SEO TV sensor system is the Computer Compatible Tape (CCT) containing the geometrically and radiometrically corrected picture data along with the relevant annotation information.

The CCT format is illustrated in Fig.3.

Depending on the number of scenes that a user has requested a tape contains(1+N) files, representing (N/2) scenes where each scene is available in two spectral bands. The first file on the tape is the Header File consisting of (1+N) records. The remaining files called picture files, consisting of (1+400) records each, contain the scene data. The two band data for each scene is given in two conseqútive files.

The header file contains a tape header and N header records, each 40 bytes long. A picture file contains one annotation record and four hundred video data records, each 400 bytes long. Except the video data records, all other records are written as 16 bit binary words. The video data record contains scene information in one byte form corresponding to a scan line. The information about the scene has been digitized so that each data point is represented by a radiance level varying from 0 to 127. The information contained in the header records and annotation records is described in Table 1 and Table 2 respectively.

TABLE 1 Header Records of CCT

Word (16 bits)	Description
	Tape Header
1	Mission Identification (1 for ISRO-SEO1)
2	Sensor Identification (1 for TV, 2 for SAMIR)
3	Number of files that follow
4-20	Unused
	Header Record
1	Year (79-81)
2	Month (1-12)
3	Date (1-31)
4	Hours (0-24)
5	Minutes (0-60)
6	Seconds (0-60)
7	Band (1-2)
8	Location grid No.latitudewise (1-40)
9	Location grid No.longitudewise (1-40)
10	No.of the archival tape on which this data is available.
11	File No.on the archival tape in which this data is available
12	Processing status. (1 for Bulk, 2 for Precision)
13	Indicates sub-status (Reserved for Future use)
14	Percentage cloud cover.
15-20	Unused.

TABLE 2 Annotation Record of CCT

1	Revolution number
2	Ground Station. 1 for Ahmedabad, 2 for SHAR.
3	Day Number Since Launch
4-6	Exposure Date (Year, Month and Day number)
7-9	Exposure Time (Hours, Minutes and Seconds)
10	Camera Identification (Band number)
11	Exposure Duration (1 for 1 ms, 2 for 1.5 ms and 3 for 2 ms)
12	Total number of bands in which data is available.
13	Latitude of subsat.point (degree part)
14	Latitude of subsat.point (minute part)
15	Longitude of subsat.point(degree part)

16 Longitude of subsat.point (minute part)
17 Direction of motion of the spacecraft w.r.t. true North in degrees (X100)
18 Satellite Altitude in Kms
19 Sun Elevation angle in degrees.
20 Sun Azimuth angle in degrees
21 Beam current (1,2,3 for low, medium, high)
22 Percentage cloud cover
23-24 Unused
25 Latitude of the format centre (degree part)
26 Latitude of the format centre (minute part)
27 Longitude of the format centre (degree part)
28 Longitude of the format centre (minute part)

Locations of 8 points on the boundary of a scene in terms of latitudes and longitudes are provided. (Refer Fig.4)

29-30 Latitude of the top right corner (degrees, minutes)
31-32 Longitude of the top right corner (degrees, minutes)
33-34 Latitude of the centre of the right edge (degrees, minutes)
35-36 Longitude of the centre of the right edge. (degrees, minutes)
37-38 Latitude of the bottom right corner (degrees, minutes)
39-40 Longitude of the bottom right corner (degrees, minutes)
41-42 Latitude of the centre of the bottom edge (degrees, minutes)
43-44 Longitude of the centre of the bottom edge (degrees, minutes)
45-46 Latitude of the bottom left corner (degrees, minutes)
47-48 Longitude of the bottom left corner (degrees, minutes)
49-50 Latitude of the centre of the left edge (degrees, minutes)
51-52 Longitude of the centre of the left edge (degrees, minutes)
53-54 Latitude of the top left corner (degrees, minutes)
55-56 Longitude of the top left corner (degrees, minutes)
57-58 Latitude of the centre of the top edge (degrees, minutes)
59-60 Longitude of the centre of the top edge (degrees, minutes)
61-62 Location grid number (latitudewise,longitudewise)
63 Processing details (Bulk = 1, Precision=2)
64 Further details about Processing/Data Quality.
65 Direction of Pass (+1 for North, -1 for south)
66 Reference document number
67 Mission Identification (1 for ISRO-SE01)
68-200 Unused.

SEO SAMIR DATA PRODUCT

The SEO SAMIR (Satellite Microwave Radiometer) data product in the form of a line printer output is illustrated in Fig.5. All the terms indicated are self-explanatory.

The first line gives the orbit number, spin number, date (day-month-year) and time (hour-minute-second) of data acquisition, the latitude and longitude of the subsatellite point in degrees and minutes and the direction of motion of the satellite in degrees measured from true North in the clockwise direction.

Then for each radiometer the beam centre of observations in terms of latitude and longitude coordinates (degrees and minutes) and the corresponding brightness temperature in degree Kelvin are given in a tabular form. The view angles for each observation for radiometers 1 and 2 and that for radiometer 3 are also indicated. The fifth sample corresponding to the view angle 180° gives the brightness temperature of the cold sky, which is used for calibration.

The characteristics of the SAMIR radiometers are listed in Table 3.

TABLE 3 Characteristics of SAMIR Radiometers.

Radiometer	1,2	3
Frequency (GHz)	19.35	22.235
Expected brightness temperature (°K) at		
i) Nadir	120-320	120-320
ii) Zenith	3	3
Temperature Sensitivity (°K)	1	1
Beamwidth (degrees)	14.02	22.86
Integration time (m.sec)	210	350
Ground Resolution in KM (with altitude = 500 KM and spin = 11 rpm)	125	200

SEO-DCP DATA PRODUCT

A sample data product of the SEO Data Collection Platform experiment is presented in Fig.6. The various parameters shown are explained below. The abbreviated terms in the heading have the following meaning associated with them.

First Line

DAYS	Day since launch
HRS	Hour of observation
MINS	Minute of observation
AIRTMP	Air Temperature
WB.TMP.	Wet bulb temperature
WINDDIR	Wind direction
WINDSPO	Wind Speed
PRESSURE	Atmospheric pressure.

Second Line

DCPHMON	DCP Health monitoring information
ENC TMP	Enclosure temperature
RFOVRFLW	Rainfall overflow
RFCOUNT	Rainfall count
SUNSHIN	Sunshine Duration

Third Line

0006	Indicates the platform number (Refer Table 4)

Fourth Line

The correspondence of this line with the heading is as indicated.

DAYS	HRS	MINS	AIRTMP	WB.TMP	WINDDIR	WINDSPD	PRESSURE
270	03	30	00 0109	01 0109	02 2756	03 0548	04 0127

Along with each sensor value there is an associated two-digit number which is the number of the corresponding sensor (Refer Table 5)

Fifth Line

This line corresponds to the second line of the heading as indicated.

DCPHMON	ENCTMP	RFOVRFLW	RFCOUNT	SUNSHIN
05 3279	06 0327	07 0080	08 7997	09 0106

Here also the two digit number indicating the number of the sensor is given for each value (Refer Table 5)

Subsequent Lines

The subsequent groups of two lines each have the same interpretation as given above for the fourth and fifth lines.

All the values given in the table are expressed in engineering units. The data set covers a duration of 24 Hours.

Unlike the other data products of SEO for sensors such as TV and SAMIR, this data product for DCP is mainly to cater to the needs of the Indian Meteorological Department. IMD has designed and constructed all the sensors used in all the data collection platforms. The data will be sent to IMD, Ahmedabad Office daily.

CONCLUSION

This paper describes the data products that will be distributed to various earth resources, and environment study specialists from data received from the Satellite for Earth Observation. This is one of a series of satellites which the Indian Space Research Organisation is planning for resources study, and the work in relation to which has created user preparedness for remote sensing applications in the country.

TABLE 4 DCP Locations

Platform No.	Location	Latitude	Longitude
1	Ahmedabad	23°04'N	72°38'E
2	SHAR	13°35'N	80°18'E
3	Delhi	28°35'N	77°12'E
4	Gulmarg	34°03'N	74°24'E
5	Jodhpur	26°18'N	73°01'E
6	Port Blair	11°40'N	92°43'E
7	Amni Devi	11°07'N	72°44'E
8	Patna	25°37'N	85°10'E

TABLE 5 Specifications of DCP Sensors

Sensor No.	Parameter	Nature	Range	Accuracy
00	Air Temperature	Analog	-10°C to 59°C	0.1°C
01	Wet Bulb Temperature	"	-10°C to 50°C	0.5°C
02	Wind Direction	"	0°C to 360°C	1°
03	Wind Speed	"	0 - 100 knots	1 knot
04	Pressure	"	100 mb over a fixed datum	1 mb
05	Power Supply Status	"		
06	Enclosure Temperature	"	0°C to 50°C	1°C
07	Rainfall Counter overflow	Digital	Overflow to 9	
08	Rainfall	"	Cumulative total 999 mm	1 mm
09	Sunshine Duration	"	0 - 19 Hrs	3 Mts.

ACKNOWLEDGMENT

The work reported here is an outcome of efforts of a team of scientists and engineers working in data product activity in ISRO, and of which the author happens to be a member. The author was helped by Mr. N.T.Unnikrishnan in preparing the manuscript. Mr. B.S.Joshi typed the manuscript.

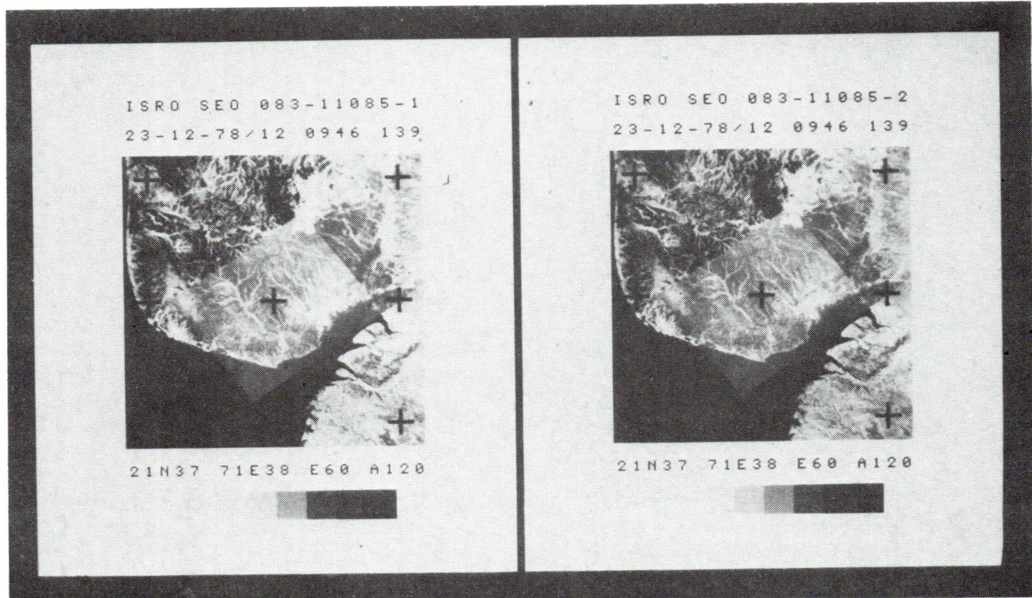

Fig.1a SEO TV Browse Product

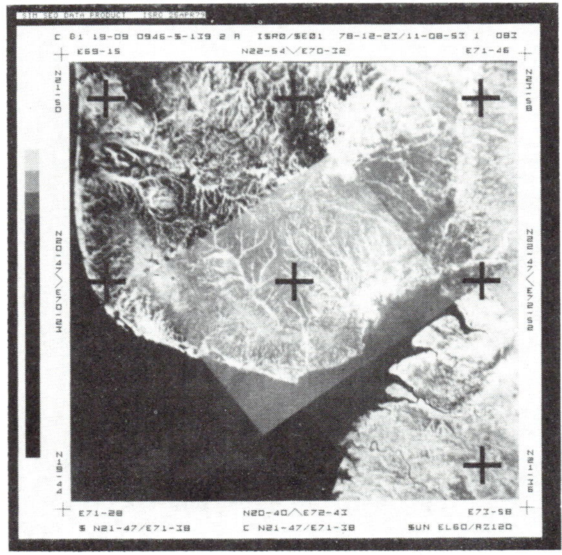

Fig. 1b. SEO TV 70 mm Product

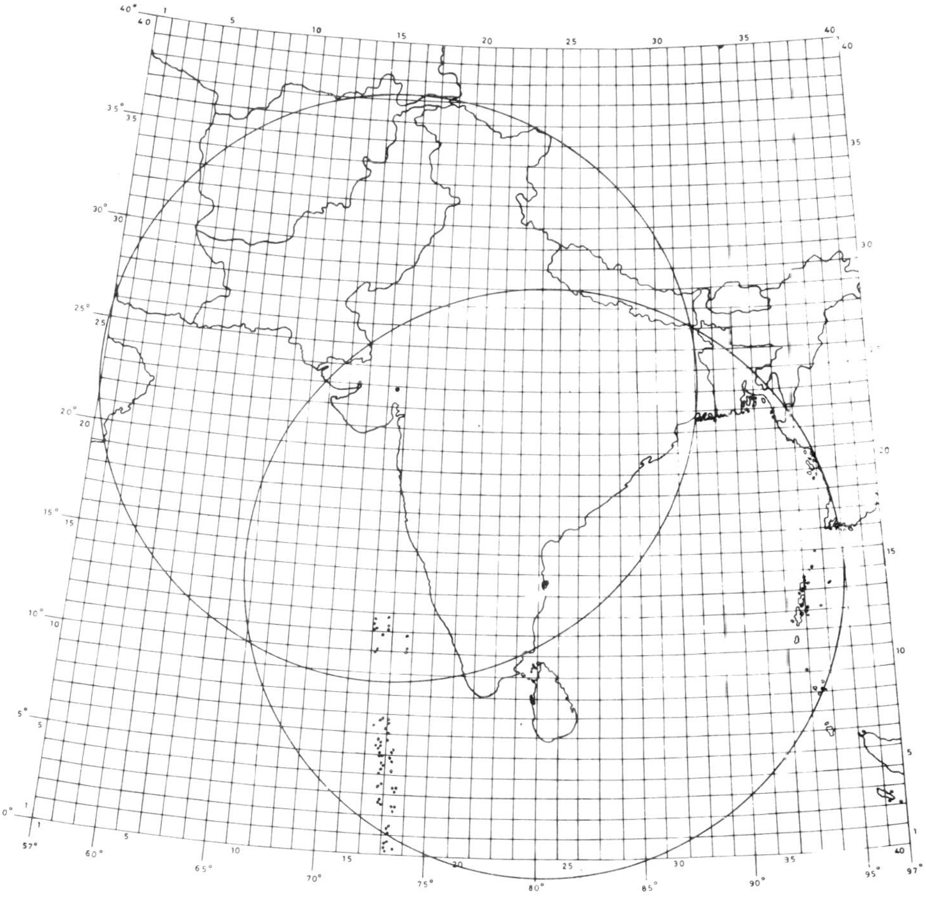

CIRCLES INDICATE VISIBILITY ZONES FOR 10° ELEVATION AND 500 Kms. ALTITUDE.

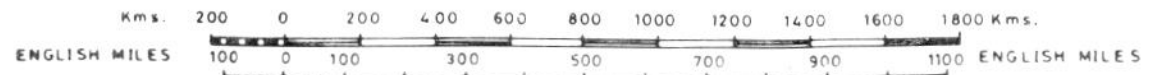

Fig. 2 Location Grid on the Map of India

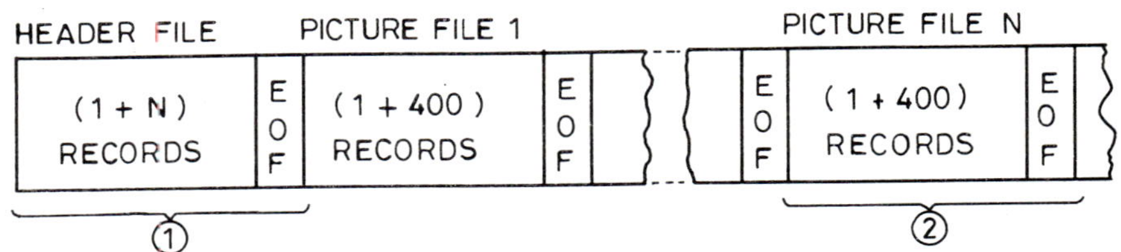

FIG. 3 SEO CCT FORMAT

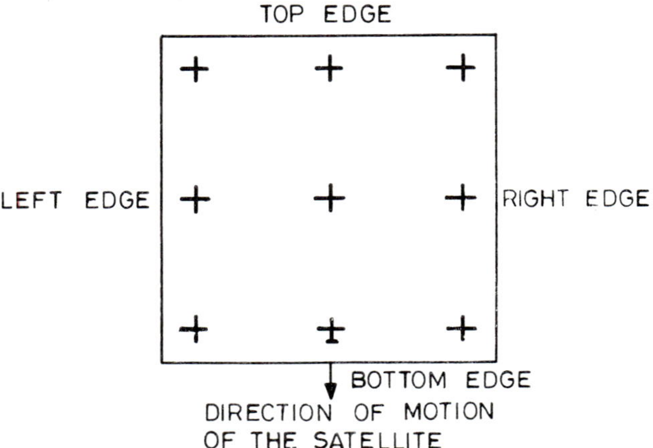

FIG. 4 ORIENTATION OF THE FACE PLATE FOR DEFINING THE ANNOTATION FORMAT

ISRO/SEO1/SAMIR/ 1

ORB. NO.	SPIN NO.	DATE(DD. MM. YY)	TIME(HH.MM.SS)	SUB SAT.(LAT.,LON)	HEAD
1	1	4-11-78	(IST) 11.25.45	10- 0,75- 0	53

SAMPLE NO.	VIEW ANGLE	RADIOMETER 1(19.35 GHz) BEAM CENTRE(LAT.,LON.)	BT(K)	RADIOMETER 2(19.35GHz) BEAM CENTRE(LAT.,LON.)	BT(K)	VIEW ANGLE	RADIOMETER 3(22·35 GHz) BEAM CENTRE(LAT.,LON.)	BT(K)
1	5.6	11- 1,75- 0	302	10-10,75-22	301	11.2	9-10,75-33	301
2	2.8	10-36,75-56	302	9-27,75-55	297	2.8	10-27,74-15	299
3	-2.8	9-36,76- 6	278	10-36,74-26	281	-2.8	9- 2,75-21	280
4	-5.6	10-53,74-48	306	9-53,74-59	305	-11.2	9-27,74-25	307
5	180.0	COLD SKY TEMP. (K)	3	COLD SKY TEMP. (K)	3	180.0	COLD SKY TEMP. (K)	3

ORB. NO.	SPIN NO.	DATE(DD. MM. YY.)	TIME(HH.MM.SS)	SUB SAT. (LAT., LON.)	HEAD
1	21	4-11-78	(IST) 11.28. 5	12- 0,78- 0	53

SAMPLE NO.	VIEW ANGLE	RADIOMETER 1(19·35 GHz) BEAM CENTRE(LAT.,LON.)	BT(K)	RADIOMETER 2(19·35 GHz) BEAM CENTRE (LAT., LON.)	BT(K)	VIEW ANGLE	RADIOMETER 3(22·35 GHz) BEAM CENTRE(LAT., LON.)	BT(K)
1	5.6	13- 1,78- 0	302	12-10,78-22	301	11·2	11-10,78-33	301
2	2.8	12-36,78-56	302	11-27,78-55	297	2.8	12-27,77-14	299
3	-2·8	11-36,79- 7	278	12-36,77-26	281	-2·8	11- 2,78-22	280
4	-5·6	12-53,77-48	306	11-53,77-59	305	-11.2	11-27,77-25	307
5	180·0	COLD SKY TEMP. (K)	3	COLD SKY TEMP. (K)	3	180·0	COLD SKY TEMP. (K)	3

ORB. NO.	SPIN NO.	DATE(DD. MM. YY)	TIME(HH. MM. SS)	SUB. SAT. (LAT., LON.)	HEAD
1	41	4-11-78	(IST) 11. 30. 25	14- 0,81- 0	53

SAMPLE NO.	VIEW ANGLE	RADIOMETER 1(19·35 GHz) BEAM CENTRE(LAT., LON.)	BT(K)	RADIOMETER 2(19·35GHz) BEAM CENTRE(LAT.,LON.)	BT(K)	VIEW ANGLE	RADIOMETER 3(22.35 GHz) BEAM CENTRE(LAT.,LON.)	BT(K)
1	5.6	15- 1,80-59	302	14-10,81-22	301	11.2	13-10,81-33	301
2	2.8	14-36,81-57	302	13 27,81-56	297	2.8	14-27,80-14	299
3	-2.8	13-36,82- 7	278	14-36,80-25	281	-2·8	13- 2,81-22	280
4	-5.6	14-53,80-48	306	13-53,80-59	305	-11.2	13-27,80-25	307
5	180·0	COLD SKY TEMP. (K)	3	COLD SKY TEMP. (K)	3	180.0	COLD SKY TEMP. (K)	3

FIG. 5 SEO SAMIR DATA PRODUCT

DAYS	HRS	MINS	AIRTMP. DCPHMON	WB. TMP. ENC TMP.	WINDDIR RFO VRFLW	WINDSPD RF COUNT	PRESSURE SUNSHIN
0006							
270	03	30	00 0109	01 0109	02 2756	03 0548	04 0127
			05 3279	06 0327	07 0080	08 7997	09 0106
270	04	30	00 0109	01 0109	02 2756	03 0548	04 0128
			05 3279	06 0327	07 0080	08 7997	09 0106
270	05	30	00 0109	01 0109	02 2756	03 0548	04 0127
			05 3279	06 0328	07 0080	08 7997	09 0106
270	06	30	00 0109	01 0109	02 2756	03 0548	04 0127
			05 3284	06 0327	07 0080	08 7997	09 0106
270	07	30	00 0109	01 0109	02 2756	03 0548	04 0127
			05 3279	06 0327	07 0080	08 7997	09 0106
270	08	31	00 0109	01 0109	02 2756	03 0548	04 0127
			05 3279	06 0327	07 0080	08 7997	09 0106
270	09	30	00 0109	01 0109	02 2756	03 0548	04 0128
			05 3279	06 0327	07 0080	08 7997	09 0106
270	10	30	00 0109	01 0111	02 2756	03 0548	04 0127
			05 3279	06 0327	07 0080	08 7997	09 0106
270	11	30	00 0109	01 0109	02 2756	03 0551	04 0128
			05 3279	06 0327	07 0080	08 7997	09 0106

FIG. 6 SEO-DCP DATA PRODUCT

MINERAL RESOURCE EXPLORATION, INVENTORY AND ASSESSMENT

W. D. Carter

U.S. Geological Survey, Reston, Virginia, U.S.A.

SUMMARY

Digital Landsat data in the form of computer compatible tapes (CCT's) are a natural supplement to existing or developing digital mineral resource inventory and geographic information systems. However, no one has yet attempted to design a total system which combines these valuable data sets. A growing body of literature among which are Abrams (1978), Abrams and others (1977), Blodget and others (1978), Bolviken and others, (1977), Lyon and Prelat (1978), Huntington and Green (1978), Schmidt (1976), Taranik (1978), and Taranik and others (1978), describe digital methods of enhancing Landsat images and multispectral analysis for geologic applications, thereby, improving rock discrimination, highlighting rock alteration zones by ratioing methods, and spectral measurement of rocks and rock products.

Mitchell and others (1977), describe a Geographic Information Retrieval and Analysis System (GIRAS) designed for handling land use and land cover data. Maps at scales of 1:250,000 and 1:100,000 were created primarily by visual interpretation of aerial photography and later digitized into computer format. In addition to land use and land cover the maps show hydrologic units, political units, census county subdivisions, Federal land ownership, State land ownership. Gaydos and Wray (1978), have demonstrated the use of digital Landsat data to create the same types of information by computer at the same scales. They have, however, eliminated the tedious step of digitizing the maps after they have been made. The Land Cover Map of the Washington, D.C. Urban Area was produced at a scale of 1:100,000 using the LARSYS III software program developed by the Laboratory for Remote Sensing (LARS) at Purdue University. These programs are available on sale to the public through the COSMIC Distribution Center, Atlanta, Georgia. The work was done on the University of Illinois ILLIAC IV high speed computer located at NASA's Ames Research Center in California. The map is subdivided into units based on a Universal Transverse Mercator (UTM) grid system. Statistical data representing each land cover class are printed by unit location on the back of the map.

Calkins and others (1978), describe CRIB, the U.S. Geological Survey's Computerized Resources Information Bank, which has been made available for public use through the computer facilities at the University of Oklahoma and the General Electric Co., U.S.A. It is available on a worldwide basis through the use of General Electric's worldwide information services network. U.S. Geological Survey Circular 755 chapter A is a manual describing CRIB as a set of variable-length records on the metallic and non-metallic mineral resources of the United States and other cooperating countries. Over 31,645 records are available in the master file. Information on mineral deposits and mineral commodities include: deposit name, location, commodity information, description of deposit, geology, production, reserves, potential reserves and references. Chapter B of the manual describes the individual data items and codes in CRIB (Keefer and Calkins, 1978).

Clark (1977), provides guidelines for international data collection and resource assessments and analyses based on experience gained in the United States, Bolivia, Turkey, and Argentina. The procedures developed have been expanded to other parts of the World through the International Geological Correlation Program Project 98 entitled "Standards for Computer Applications in Resource Studies." The initial accomplishments of the project have been described by Cargill and others (1977).

Leaders of IGCP Project 98 and Project 143 entitled "Remote Sensing and Mineral Exploration" have agreed to work together to determine where their activities are mutually supportive. The two most obvious areas of common interest that have thus far been defined is 1) to add the number of the best available Landsat images to the CRIB location information so that the users of CRIB can obtain an image if they should wish to study the area remotely to consider and/or design an exploration strategy, 2) the second is to add Landsat image analysis report references to either the "Exploration and Development" record or to the General References of CRIB. This will enable the users of CRIB to determine the results of earlier examinations using remotely sensed data. An initial source for reference information is contained in IGCP Project 143 Bibliography on Remote Sensing and Mineral Exploration which is computerized on the U.S. Geological Survey's MULTICS system.

REFERENCES

Abrams, M. J., 1978 (Computer Image Processing - Geologic Applications:) Jet Propulsion Laboratory, Pasadena, California Publication,78-34, 27p.

Abrams, J. J., Ashley, R. P., Rowan, L. C., Goetz, A. F. H. and Kahle, A. B., 1977, (Mapping hydrothermal alteration in the Cuprite mining district, Nevada, using aircraft scanner images for the spectral region 0.46 to 2.36 um: Geology vol. 5, no. 16, p. 713-718.

Blodget, H. W., Gunther, F. J., and Podwysocki, M. H., 1978, (Discrimination of Rock Classes and Alteration Products in Southwestern Saudi Arabia with Computer Enhanced Landsat Data:) NASA Technical Paper 1327, 34p.

Calkins, J. A., Keefer, E. K., Ofsharick, R. A., Mason, G. T., Tracy, Patricia and Atkins, Mary, 1978 (Description of CRIB, the GIPSY Retrieval Mechanism, and the Interface to the General Electric MARK III Service:) U.S. Geological Survey Circular 755-A, 49p.

Cargill, S. M., Meyer, R. F., Picklyk, D. D. and Urquidi, F., 1977, (Summary of Resource Assessment Methods Resulting from the International Geological Correlation Program Project 98:) Mathematical Geology, vol. 9, No. 3, p. 211-220.

Chavez, P. S. Jr., O'Connor, J. T., McMaken, D. K., Eliason, Eric, 1979, (Digital Image Processing Techniques of Integrated Images and Non-Image Data Sets:) Proc. 13th Int. Symposium on Remote Sensing of the Environment. Preprint 16p.

Clark, A. L., 1977, (Guidelines for International Data Collection and Resource Assessments and Analyses:) Mathematical Geology, vol. 9, no. 3, p. 221-233.

Gaydoes, Leonard and Wray, J. R., 1978, (Folio of Land Use in the Washington, D.C. Urban Area:) U.S. Geological Survey Map I-858E.

Keefer, E. K. and Calkins, J. A., 1978, (Description of Individual Data Items and Codes in CRIB:) U.S. Geological Survey Circular 755-B, 32p.

Mitchell, W. B., Guptill, S. C., Anderson, K. E., Fegeas, R. G., and Hallam, C. A., 1977, (GIRAS: A Geographic Information Retrieval and Analysis System for Handling Land Use and Land Cover Data:) U.S. Geological Survey Professional Paper 1059, 16p.

Podysocki, M. H., Gunther, F. J. and Blodget, H. W., 1977, (Discrimination of Rock and Soil Types by Digital Analysis of Landsat Data:) NASA Goddard Spaceflight Center, Greenbelt, Md., Preprint X-923-77-17, 37p.

Taranik, J. V., 1978, (Principles of Computer Processing of Landsat Data for Geologic Applications:) U.S. Geological Survey Open File Report 78-117, 50p.

Taranik, J. V., Reynolds, C. D., Sheehan, C. A. and Carter, W. D., 1978, (Targeting Exploration for Nickel Laterites in Indonesia with Landsat Data:) 12th International Symposium on Remote Sensing of the Environment, Manila, Phil., Proc., p. 1037-1051.

WORKSHOP EXERCISE FOCUSED ON STRUCTURAL GEOLOGY AND MINERAL RESOURCES OF KARNATAKA STATE, INDIA: EXPLANATORY NOTE

W. D. Carter

U.S. Geological Survey, Reston, Virginia, U.S.A.

Two hours of our program were devoted to a workshop exercise in which thirty of the participants were given black and white, Band 7 multispectral scanner (MSS) images to analyze visually on light tables supplied by the Bangalore office of the Geological Survey of India (GSI). The Survey also provided overlay material, colored marking pens, and metric scales. The Landsat images were reproduced and supplied by the Indian Space Research Organization (ISRO).

The main purpose of the exercise was to enable the participants to become familiar with the terrain, bedrock, drainage patterns and structural geology of the Mysore area by studying the Landsat images so that they could better evaluate the results of the paper on the region to be presented by Dr. J.G. Krishnamurthy of GSI, which follows. The program organizers used the opportunity to view, review and discuss the work of each of the participants on an informal basis to see how fast they worked and the degree of detail they were able to interpret. While we did not attempt to statistically evaluate these results, we believe that the exercise was beneficial in breaking down the artificial barriers of more formal meetings and that it prepared them for the later discussions. The participants were provided with a product covering the area of their visit which they could take home with them and use in demonstrations to others.

The workshop also provided the opportunity to distribute and discuss instructional papers and reprints that were provided to each of the participants by the EROS Program of the U.S. Geological Survey and which were reproduced by ISRO. These papers included the following:

1) Integration of Geological Remote Sensing Techniques in subsurface Analysis by J.V. Taranik and C.M. Trautwein, U.S. Geological Survey Open File Report 76-402, 60 p. (1976).

2) Characteristics of the Landsat Multispectral Data System, by J.V. Taranik, U.S. Geological Survey Open File Report 78-187, 76p. (1978).

3) Principles of Computer Processing of Landsat Data for Geologic Applications, by J.V. Taranik, U.S. Geological Survey Open File Report 78-117, 50 p.

4) Reprint of American Scientist, vol. 63, no. 4, p. 393-403. "Application of satellites to geologic exploration" by L.C. Rowan (1975).

5) Reprint of Geology, vol 5, no. 12, p. 713-718, December 1977. "Mapping of hydrothermal alteration in the Cuprite mining district, Nevada, using aircraft scanner images for the spectral region 0.46 to 2.36 m." by M.J. Abrams, R.P. Ashley, L.C. Rowan, A.F.H. Goetz and A.B. Kahle.

On display were copies of the following references:

1) "ERTS-1: A New Window on Our Planet." Edited by R.S. Williams Jr., and W.D. Carter, U.S. Geological Survey Professional Paper 929, 362 p. (1976).

2)"Proceedings of the First Annual W.T. Pecora Memorial Symposium, October 1975, Sioux Falls, South Dakota" edited by P.W. Woll and W.A. Fischer; U.S. Geological Survey Professional Paper 1015, 370 p. (1977).

In addition the Geological Survey of India displayed national and regional geologic maps which included a preliminary Lineament Map of India. Various examples of enlarged color composite Landsat images processed by the Indian Space Research Organization and examples of color enhanced and ratio images of India processed by J.F. Huntington of CSIRO, Australia were also displayed, viewed and discussed.

W.D. Carter presented preliminary results of interactive computer analysis of the Mysore Landsat image (E-2738-04185) which used parallelepiped programs to discriminate between and measure the reflectance spectra of the three major rock types and two vegetation types in the area. An attempt was also made to measure the spectra of a known chromite-bearing ultramafic rock in the Arsikere region within the scene. These measurements were then compared with other similar localities in the region and an exploration target map was developed which will be field checked by the Geological Survey of India. A complete set of slides and histograms which explain the process used was given to the Bangalore office of the Indian Geological Survey.

GEOLOGICAL GROUND-TRUTHS AND LANDSAT IMAGERY INTERPRETATION FOR PARTS OF KARNATAKA STATE (INDIA)

J. G. Krishnamurty, K. P. Gopalakrishnan and T. V. Ramachandran

Airborne Mineral Surveys and Exploration Wing, Geological Survey of India, Bangalore 560041, India

ABSTRACT

The purpose of this paper is to evaluate the geological and structural concepts relating to "Dharwar" [1] geology and to highlight some of the salient observations by the authors from the Landsat images for a part of southern Karnataka State between Mangalore and Kolar in southern India.

Visual study of the LANDSAT-I images of the frames 154-051 (Bangalore area), 155-051 (Mysore area) and 156-051 (Kudremukh area) of southern Karnataka using diapositives of the three images, 70 mm negatives and diapositives and colour composite overlays (Diazo overlays) and prints and correlation with the ground geological, geochemical, geophysical and exploratory data and airborne geophysical data by the authors resulted in several significant observations like i) the possible continuity of the Holenarsipur schist belt to Bababudan Iron-ore through a highly folded migmatite zone, (ii) the delineation of a schistose band surrounding the Arsikere granite, (iii) the bounding of Closepet granite zone by two well defined N-S trending lineaments, (iv) recognition of two circular zones defined by a high drainage density and semi-annular and radial patterns of drainage around Nuggihalli and NE of Mysore, indicating the possibility of the presence of zones of reactivated basement, (v) the straight courses of the rivers Cauvery, Vedavati and Yagachi indicating their control along some major lineaments, (vi) the clockwise rotation of the dominant structural trend from NNW-SSE in the west (Frame 156-051) to ENE-WSW in the east (Frame 154-051) and (vii) the identification of a fold with a NW-SE trending axial trace near Mercara.

INTRODUCTION

The paper attempts to draw out the salient features of Karnataka geology, derived from Landsat interpretation, as the frames selected for study encompass the well-known areas like Bababudan (13°30':75°45') iron ore belts and the other schist belts in southern Karnataka, including Kolar (13°17':78°08') schist belt in the east, all of which bristle with innumerable problems, that provide the basis for debates on many platforms.

GEOLOGICAL AND STRUCTURAL SETUP OF THE AREA

Regional Geology:

Newbold was the pioneering geologist, who described the geology of some regions in Karnataka State and separated the schists from the gneisses. He also recorded the instances of intrusion of gneiss into schists.

Bruce Foote [1] [2] [3] applied the term "Dharwar System" to designate all crystalline schists of volcano-sedimentary origin and elongated synclinal bands resting unconformably on a basement complex of granitoid gneiss. He had assumed the sedimentary origin of Dharwar schists.

Holland [4] [5] [6] was, however, of the opinion that the Dharwar rocks are not younger than the peninsular gneisses.

Smeeth [7] considered Dharwars to be entirely of igneous origin and divided the same into a lower hornblendic division and an upper chloritic division.

B.Rama Rao [8] classified the Dharwars into three divisions-upper, middle and lower, on the basis of intervention of two series of conglomerates. The lower division is completely volcanic, whereas the middle and upper are mainly sedimentary type.

Coming to the recent times, Nautiyal [9] , Radhakrishna [10] and Swami Nath et al [11] have expressed separately that the gneisses form basement to the schists. Radhakrishna [12] and Swami Nath et al [13] have opined that the gneisses intrude the high grade schists, but form a basement to a number of younger schist belts.

S.V.P. Iyengar [14] suggested a four-fold classification of the Dharwars based on an attempt to correlate the various belts with the help of common horizons. According to him, the oldest group is 'Mercara (12°25':75°44') Group', comprising the biotite-kyanite schists and the associated charnockite-leptynite of Mercara. The 'Bababudan Group' overlies Kalaspura (13°17':75°56') granite-gneiss with an angular unconformity. The Bababudan Group is overlain unconformably by 'Vanivilas Group', which in turn, is overlain unconformably by rocks of the 'Ranibennur (14°37':75°38') Group'.

Naqvi and Hussain [15] indicated the presence of two sequences in the Dharwar Super Group, one older than 2600 m.y. and another younger. The former was called 'true greenstone belts' and the latter 'geosynclinal piles' (Naqvi et al [16]).

Swami Nath et al [13] recognised two distinct belts west and east of the Closepet (12°43':77°19') granite based on lithostratigraphic correlation of almost all the schist belts of southern Karnataka. According to them, the western belt is characterised by the mature sediment-dominated greenstone belts with subordinate volcanism and intermediate pressure metamorphism and the eastern block is marked by the volcanics-dominated greenstone belts and low pressure metamorphism.

Based on the available geochronological data on the gneissic complex and younger granites, a major tectonometamorphic episode of Dharwar Super Group ranging from 2600 m.y. and 2300 m.y. has been recognised (Swami Nath et al [13]), corresponding to which a major event of granitic activity was identified in the gneissic complex, between 2500-2600 m.y. (Venkatasubramanian [17]). They are of the view that this granitic event can be correlated with the emplacement of Arsikere (13°19':76°15') and Banavar (13°24': 76°10') plutons (isochron 2620 m.y.). They are of the opinion that the Closepet Suite of granites is a result of several granitic events, culminating in a major peak at 2100-2300 m.y. Chamundi (12°16':76°40') granite, according to them, is the youngest granitic event in the craton (790 ± 60 m.y.).

The Dharwar craton is partly surrounded by the Eastern Ghats mobile belt, which is characterised by high grade metamorphic rocks such as charnockites and khondalites. Pichamuthu [18] considers the high grade rocks as 'exposed co-eval roots of lower grade volcano-sedimentary-granite suites'.

Ramiengar <u>et al</u> [19] are of the opinion that the high grade event of 2900-3000 m.y. (which has affected the gneiss and the enclaves) was responsible for the formation of charnockites by a process of dehydration at deeper levels.

<u>Structure:</u>

The structural information available for Karnataka State is mainly for the schistose belts and very little published data are available for the Peninsular Gneissic Complex.

The general trend of the schistose belts varies from NNW-SSE to NW-SE in the middle and northern parts of Karnataka, while it swings to N-S, changing to SW and WSW in the southern part.

Iyengar [14] has recorded an earlier WNW-ESE fold trend along the west coast and in the Bababudan region, which is also noticed in the southern part of the Ranibennur-Shimoga (13°55':75°35') schist belt. According to him, this older belt is truncated by later NNW trend. He also mentions a third NE-SW trend, recognisable in parts of Ranibennur-Shimoga schist belt, the Bababudan belt and the Gadag (15°26':75°39')-Chitradurga (14°14':76°23')-Seringapatam (12°25':76°41') belts. Iyengar [14] considers the Dharwar exposures in south Mysore as a series of a detached synclines infolded into the basement. According to him, a thrust running close to the southern border of Mysore (12°19':76°38') separates the charnockite region from the Dharwar region.

The regional structure in Karnataka craton is held to be a 'Synclinorium plunging northward' (Ramiengar <u>et al</u> [19]).

Recent structural studies in the linear supracrustal belts and adjacent Peninsular Gneiss in the areas of Sargur, Holenarsipur (12°48':76°15') and Chitradurga by Chadwick <u>et al</u> [20] have revealed three phases of deformation in both the older Sargur Group and the younger Dharwar Super Group.

BRIEF ACCOUNT ON MINERAL OCCURRENCES AND ON GEOCHEMICAL GEOPHYSICAL AND EXPLORATORY DATA

Karnataka is one of the best states of India in mineral resources, where a variety of important mineral occurrences have been located, mainly in the high grade schists, while only a few minerals like quartz, feldspar, mica, kaolin and some rare-earth minerals are found in the pegmatites traversing the gneissic region.

Brief account, mineral-wise [21] , is as follows:

1. Precious Metals (Gold, Silver and Platinum)

Karnataka is the only state in India that has been producing gold since 1880. Production comes from the Kolar gold field and the Hutti (16°15':76°38') gold field. Other gold fields in the state, namely Gadag, Bellara (13°35':76°35') and Honnali (14°14':75°39') have also produced gold. The production of gold has steadily been declining in recent years. However integrated surveys and detailed large scale mapping followed by drilling by GSI during the last few years have indicated additional reserves of gold in the Kolar gold fields.

The latest estimate of gold in the working mines are:
(M.Ziauddin and S.Narayanaswami [22]).
Kolar Gold Field:

Payable	9,470 Kgs.
Low grade	12,870 Kgs.
Hutti Gold Field	690 Kgs.

At present the tenor of gold mined is 6-7 gms/ton. Silver is obtained only as a by-product from the smelting of gold. The A.M.S.E. Wing, GSI has recently undertaken detailed studies of ultramafic belts in south Karnataka (viz.Banasandra (13°16': 76°40') Haradasanahalli (13°38':76°40'), Nuggihalli (13°01': 76°29') and Sindhuvalli (12°12':76°38')- Kadakola (12°11':76°40')) with a view to locating platinoid group of minerals.

2. Non-Ferrous Minerals

Copper

Ingaldhalu (14°11':76°27')

Ancient workings for copper ore are noticed on the western flanks of Belligudda hill, near Ingaldhalu about 8 Kms. SSE of Chitradurga. The Department of Geology and Mines, Karnataka, drilled 5440 metres in 45 bore holes, which indicated a reserve of 1,000,000 tonnes of 1.1% copper. The deposit is presently being mined by the Chitradurga Copper Company.

The GSI has carried out detailed mapping, geochemical and geophysical prospecting, pitting and drilling in the Kunchiganahalu (14°12':76°27') section of the Ingaldhalu sulphide zone. A reserve of 0.25 million tonnes of 1.46% copper has been estimated.

Kalyadi (13°14':76°09'):

GSI has carried out detailed mapping, geochemical and geophysical prospecting followed by diamond drilling of 5000 metres in 30 bore holes (includes 4 bore holes drilled by geology and Mines Department of Karnataka State).

Drilling has indicated a reserve of about 1.04 million tonnes of 1.32% average grade of copper. This is also being mined by Chitradurga Copper Company. Underground development has proved 1 million tonnes of ore with an average grade of 0.86 % copper.

Thinthini (16°26':76°46'):

This prospect is in Gulburga (17°19':76°50') district, where pyrite-arsenopyrite-chalcopyrite mineralisation is localised along a major fault zone in a meta-gabbro body. The drill-indicated reserve of this prospect is 6 million tonnes of 0.8 % copper.

One of the noteworthy discoveries made, using airborne geophysical data in an otherwise gneissic terrain, is in Aladahalli (13°08': 76°22') (Hassan district) area, where integrated ground survey followed by drilling indicated the presence of sulphides occuring in a high-grade aluminous schist. This has opened up tremendous scope in searching for similar occurrences within the monotonous gneissic terrain by resorting to rigid and systematic studies in certain similar environments. Devagandanahalli (13°19':75°56')-Kalaspura (Chikmagalur (13°19':75°47') district) Sowanahalli (12°08':76°48') (Mysore district) and Machnur (Raichur district). Mothimakki (14°51':74°30') (North Kanara district) and Hasakere and Medakeripura (14°14':76°27') (Chitradurga district) are the other copper prospects, where integrated surveys (large scale geological mapping, geochemical and geophysical surveys) have been followed by exploratory drilling. Kaiga (North Kanara) is being explored by drilling after the completion of integrated surveys.

Zinc Ore:

In Kunchiganahalu, Chitradurga district, exploratory drilling over geochemical anomalies revealed an ore body of 1 metre width with zinc and copper mineralisation.

Tin Ores:

The tin occurrences of Karnataka near Dambal (15°18':75°47') (Dharwar district) and Chikkannanahalli (14°23':76°22') (Chitradurga district) are of poor quality.

Bauxite:

Bauxite occurrences of Karnataka are mainly in Belgaum (15°50': 74°32'), Chikmagalur and Chitradurga districts, of which the Belgaum deposits are commercially significant. The Belgaum plant of the Indian aluminium company has an annual capacity of 30,000 tonnes of aluminium.

3. Ferrous Metal Deposits

Karnataka ranks fourth among the iron-ore producing states and has a total reserve of about 1000 million tonnes (1977). The iron ore deposits of Karnataka fall into two groups. The first type formed by leaching and cementation is essentially of hematite with cappings of limonite and goethite. These form thick caps on ferruginous shales, banded magnetite quartzite and banded hematite quartzite and stand out as prominent ridges.

The ores are mostly laterite/massive/or powdery blue dust. The major hematite deposits of Karnataka are in Sandur basin (Donimalai (15°05':76°37'), Kumaraswami (15°00':76°34'), Ramandurg (15°08':76°28'), Kanevehalli (15°02':76°30'), Devadari (15°00':76°35') and Timmapanagudi (15°08':76°33')) of Bellary district (estimated reserve 1200 million tonnes, proved reserve 583 million tonnes); Chikmagalur district (Attigundi (13°26': 75°44') Galikere, Gundikan (13°29':75°44'), Rudragiri (13°24': 75°41') and Jensarigudda of Bababudan ranges, which have vast potentialities of both hematite and magnetite), North Kanara, Shimoga, Chitradurga and Tumkur (13°20':77°06') districts. The deposits of Bababudan ranges (Chikmagalur district) supply iron ore to the Mysore Iron and Steel Works at Bhadravathi. The second type of deposits are the banded magnetite deposits of Bababudan-Kudremukha (13°21':75°15')-Kodachadri belt. Ground magnetic survey was carried out in parts of Bababudan ranges by GSI.

Titaniferous magnetite deposits occur as lenticular bands in ultrabasic rocks and amphibolites in Nuggihalli schist belt. Exploration for vanadium in the vanadiferous magnetite deposit in Masanikere, Shimoga district, is in progress.

Manganese deposits

The manganese deposits of Karnataka are of the lateritoid type or occur as local replacements within shales and limestones. The common ore minerals are pyrolusite, psilomelane and wad with subordinate manganite and braunite. The manganese deposits are located in North Kanara, Belgaum, Shimoga, Bellary and Chitradurga districts.

Chromite deposits

Important chromite-bearing ultrabasic belts of Karnataka area are those of Nuggihalli-Arsikere (Hassan district) and Nanjangud (12°07':76°41')-Mysore (Mysore district). The major occurrences are in Bairapur (13°09':76°12'), Bhaktarhalli (13°18':75°59'), Jambur (12°30':75°49') Chickannanahalli, Tagadur (13°14':75°45') etc. (Nuggihalli belt), Sindhuvalli, Talur (12°12':76°36') and Doddakatur (12°10':76°37') etc, (Mysore district). Geophysical surveys were also carried out in these belts by GSI.

4. Non-Metallic Deposits

Karnataka has extensive deposits of sedimentary limestone (Bhima and Kaladgi (16°12':75°30') limestone) and some large pockets of crystalline type (Kalinadi, Supa Dandeli belts and Chikkanayakanhalli belts). Exploratory work for locating flux grade limestone deposits was carried out in Belgaum, Bijapur (16°50':75°44'), Gulbarga (17°19':76°50') and North Kanara districts, which indicated a large reserve of flux and cement grade limestone.

Asbestos

Production of amphibole group of asbestos (anthophyllite and tremolite) comes from the Holenarsipur schist belt, where it is associated with serpentinite and tremolite-actinolite schist. No major occurrence of chrysotile asbestos is reported from Karnataka.

5. Gem Stones

Ruby

Occurrence of ruby associated with kyanite was reported from Chikmagalur district. Occurrences of corundum ruby are known from Channapatna (12°30':77°30') (Bangalore district) and Pavagada (14°15':77°15') (Tumkur district). Occasional occurrences of ruby and star ruby have been reported from parts of Mysore district.

SOME SALIENT POINTS ABOUT THE LANDSAT IMAGES SELECTED FOR INTERPRETATION

The images for the Landsat frames 154-051 (Bangalore area), 155-051 (Mysore area) and 156-051 (Mangalore area) covering parts of southern Karnataka State from Kolar in the east to Mangalore (12°51':74°50') in the west were studied. The images studied with their dates and cloud cover are given below:

Landsat Frame	Image No.	Date	Cloud cover %
154-051	E-1219-04392	27-2-73	0
155-051	E-1202-04445	10-2-73	0
156-051	E-1203-04504	11-2-73	0
	E-2035-04273 (NASA Image)	26-2-75	10

Diapositives of the three images in bands 4,5,6 and 7 (23 cm format 1:1,000,000 scale), 70 mm negatives and diapositives (1:3.369 million scale) and colour composite overlays (Diazo overlays) were used for the interpretations.

Frame 154-051 includes Clospet suite of granites (in part), Nandi (13°22':77°43') and other unclassified granites, charnockites, Kolar schist belt and the Peninsular Gneissic Complex and located administratively in three states namely Karnataka (western part), Andhra Pradesh (North-eastern part) and Tamil Nadu State (South-eastern part). Charnockites stand out conspicuously by a distinct relief in the southern part, occuring as enclaves within the gneiss or enveloping gneissic pockets. The granites occur as inselbergs within the gneissic country, as seen in Closepet suite of granites or as continuous hill ranges as in the eastern part. Kolar schist belt known for its association of Gold Fields runs along a NNW-SSE trend in the central part, bifurcating into two shoots at the south near Mallappakonda (12°45':78°10').

The principal rivers draining the area of the Landsat frame include Jayamangali, Papagni, Ponnaiyar and Palar rivers. The first two flow north, while the last two drain due south.

Frame 155-051 comprises parts of Dharwar Super-Group, part of Sargur group, Banavar, Arsikere and Chamundi granites, Closepet suite of granites, charnockites and Peninsular gneiss. Charnockite exposed in the south-western and south-eastern parts has the highest relief (maximum altitude 1713 M) and the granites occur as inselbergs within the gneissic country and are marked by distinct relief. The greenstone belts comprising the Dharwar Super-Group and Sargur Group run in the central part of the frame more or less along a NNW-SSE direction in an arcuate fashion with its concave side to the west. The Peninsular gneissic terrain is marked by flat or gently rolling topography.

The principal rivers draining the area of the frame include Cauvery river, running along a WNW-ESE lineament, Vedavati river (running along a NE-SW lineament), Hemavati river, Lakshmantirtha river, Kabbani river, Shimsha river and Yagachi river. Of these Hamavati, Lakshmantirtha and Kabbani drain into K.R.reservoir located at the trijunction of Cauvery, Lakshmantirtha and Hemavati rivers.

Frame 156-051 covers the Arabian Sea, the coastal belt, the NNW-SSE trending Western Ghats (which runs diagonally across the frame) and the inland plateau to the east. The greater part of the frame, except the coastal tract, comprising the recent formations is made up of Peninsular Gneiss and charnockites that occupy the southern part. The Dharwar Super-Group is represented by the Agumbe (13°31':75°05')-Kudremukh belt and Bababudan belt in the northern part.

The prominent hill ranges in the frame include the sickle-shaped Bababudan ranges, the horse-shoe shaped Kudremukh iron-ore ranges and the lofty hill ranges, capped by charnockites and Peninsular gneisses in the southern part. The highest peak of the frame is 1892 M, capping the Kudremukh ranges. The principal rivers draining the frame comprise Tunga and Bhadra (flowing north-east to unite to form Tungabhadra north-east of Shimoga), Yagachi and Hemavati (flowing south-east), Netravati (Flowing west) and Cauvery, which originates in the Western Ghats and flows east. The frame is located administratively in two states, Kerala (southern part) and Karnataka.

LANDSAT IMAGE INTERPRETATION AND BROAD RESULTS

The significant observations made during the study of the images are discussed below:

1. Rocks of Dharwar Super-Group (except meta-sediments) and Sargur Group do not show any distinct tonal density variations to be marked separately and hence both of them have been grouped together under the term 'Greenstone belt'.

2. The greenstone belts are clearly brought out by the dull-grey to dark grey tones and relatively poorer drainage density within them. Their tone is darker than the gneiss, but lighter compared

to charnockites and granites. Within the greenstone belts, the lighter tone represents the metasediments and the darker tone represents the meta-volcanic suite, the latter generally fringing the greenstone belts.

3. The Agumbe-Kudremukh greenstone belt can be distinguished from the surrounding granites and gneisses by their grey tone and their horse-shoe shape at the southern extremity, besides distinct relief.

4. The Bababudan belt is identified by its sickle-shaped outline, grey tone, poorer drainage density and thick vegetation cover. Folding in the belt is brought out by the resistant ferruginous quartzite bands.

5. The Holenarsipur belt is traced continously to the Bababudan belt, as indicated in the aerial photographs of this region, whereas the geological map of Karnataka State (1971) does not show the continuity. The schists in between are highly folded, as suggested by the zig-zags in the boundaries of the schist belt.

6. A zone of schist is traced from the northern border of Arsikere granite and connects up with main greenstone belt north of Karadi (13°16':76°34'). Outcrops of schist with some gossan zones have been recently identified east of Arsikere granite (Palaniappan, Personal communication).

7. A schist belt, comprising two NNW-SSE trending lensoid bodies, shifted by an ENE-WSW fault, appears to run from the western margin of Hessarghatta (13°08':77°29') tank northwards for a distance of 25 Km.

8. A NW-SE aligned schist belt is identified in the region of Kunigal (13°01':77°02') ridge.

9. The charnockites are characterised by dark grey tones, distanct relief, rugged topography and relative rarity of drainage. The high grade gneisses bordering the charnockites to the north are marked by light grey to greyish-white tones and a very high drainage density.

10. The Banavar granites show up as a group of three small granite bodies instead of a single body, as shown on the geological map and the Arsikere granite as a single large body. They have a dark tone, coarse texture and distinct relief.

11. The Closepet suite of granites appears as a number of inselbergs and is well-marked by two N-S bounding lineaments defined by discontinuous dolerite dykes.

12. The Nandi granites, Arsikere, Banavar and Chamundi granites are all similar in tone and texture to the Closepet suite of granites.

13. The Peninsular gneissic terrain is marked by whitish to greyish white tones, poor relief and coarse dendritic pattern of drainage.

14. A glance at the drainage maps of the three frames reveals that drainage is very prominent in the gneissic terrain with dendritic patterns in general with occasional radial, annular, semi-annular and rectangular patterns, while it is relatively less in the charnockites, granites and greenstone belts. In granites and charnockites, it is mostly fracture-controlled and of rectangular type, while trellis pattern of drainage is also noticed occasionally. The drainage map of Landsat frame 155-051 shows two distinct circular patterns around Nuggihalli and north-east of Mysore. They cover an area of 1960 sq.Km. and 1250 sq.Km. respectively and are characterised by radial and semi-annular drainage patterns with high drainage density. Both of them occur in the gneissic country and in the former, Nuggihalli schist belt occurs as a NNW-SSE trending linear body. These might represent zones of reactivated basement.

15. An ENE-WSW trending water shed zone, coinciding with an axis of culmination (Narayanaswami [23]), is observed north of Nandi in Bangalore scene and extending westwards into Mysore scene. The swinging of the granite inselbergs from the NNW-SSE trend to ENE-WSW in this zone suggests that this zone may be an axis of structural disturbance.

16. Several photo-linear features, identified by straight and curvilinear nature of the drainage courses, lithological boundaries, fractures, faults, vegetation and dykes have been picked from Landsat images. Some of the 'Photo-lineaments' have been found to truncate the greenstone belts and also to shift the charnockite boundaries, whereas some appear to run along prominent rivers, like Vedavati, Cauvery and Yagachi.

A study of the photo-lineaments of the three frames reveals threeprincipal trends, namely (i) WNW-ESE to E-W, (ii) N-S to NNW-SSE to NW-SE and (iii) ENE-WSW to NE-SW to NNE-SSW. They probably represent the three folding trends WNW-ESE, NNW-SSE to NW-SE and NE-SW in the Dharwar craton (Iyengar [25]). Of these, the WNW-ESE trends are mostly seen in the western frame (156-051), whereas they are less conspicuous in the other frames. The youngest structural trend NE-SW has manifestation in all the three frames and is the most ubiquitous set.

Though some of these photo-lineaments correspond to the already mapped faults, like the faults along the Vedavati and the Cauvery, the significance of many of them is yet to be understood and requires large amounts of ground truth.

The polar diagrams for the three frames, prepared from the photolinear data, show a clockwise rotation of the dominant structural trend from NNW-SSE in the west (Frame 156-051) to ENE-WSW in the east (Frame 154-051).

17. The lineaments interpreted from the images appear to be related to and have been controlled by the tectonic history of the Peninsular India, since the major lineament directions are found to be parallel and sympathetic to the major tectonic frame work and evolution of the Peninsular shield.

18. The NW-SE lineaments are the most dominant and second in importance are the ENE-WSW trending lineaments. The mineral occurrences are primarily confined to NW-SE lineaments along greenstone belts. Chromite, asbestos and magnesite are localised in the ultramafic bodies within the greenstone belts.

19. A fold with axial trace NW-SE is recognised in the charnockites north of Mercara and several E-W trend lines are seen in the charnockites west of this fold. These trend lines might represent the foliation patterns in the charnockites.

20. The outcrop configuration of the charnockite lenses in Bangalore scene suggests the presence of a number of macroscopic folds with their axial traces trending NW-SE, similar to the fold observed south of Mercara in Mysore scene.

ACKNOWLEDGEMENTS

The authors express their grateful thanks to Shri V.S.Krishnaswamy, Director-General, GSI, for the permission accorded to prepare and present this introductory paper at the COSPAR 79 Workshop inaugural session in June 1979, and to Dr.S.V.P.Iyengar, Deputy Director General, Geological Survey of India, for a critical study of the manuscript and for very valuable suggestions offered. They are greatly benefited by the discussions they have had with S/Shri B.Srikantan, B.N.Jayaram, A.V.Ramachandra, M.Ramakrishnan, M.N.Viswanatha and many other officers of GSI on various aspects of Karnataka geology and to Dr.R.Vaidyanathan, Andhra University, who had offered critical suggestions on certain geological and geomorphological aspects. The receipt of geological, geochemical and geophysical data for parts of Karnataka from Southern Region, GSI and from the Directors of Karnataka Circles (N & S), GSI is sincerely acknowledged.

REFERENCES

1. R.Bruce Foote, Rec.Geol.Surv.India,Vol.19(2), p.98-99 (1886).
2. R.Bruce Foote, Rec.Geol.Surv.India,Vol.21(2), p.40-56 (1888).
3. R.Bruce Foote, Mem.Geol.Surv.India,Vol.25(1), p.1-218 (1895).
4. T.H.Holland, Mem.Geol.Surv.India,Vol.28, p.119-249 (1900).
5. T.H.Holland, Mem.Geol.Surv.India,Vol.33(1), p.74-81 (1902).
6. T.H.Holland, Geology-Imperial Gazetteer of India,Vol.1(2), p.50-103 (1909).
7. W.F.Smeeth, Dept.Mines and Geol.Mysore State, Bull.,V.6, p.1-21 (1916).
8. B.Rama Rao, Rec.Mys.Geol.Dept.Vol.34, p.22-31 (1936).
9. S.P.Nautiyal, 53rd Indian Science Congress Proc.pt.2, p.1-14 (1966).
10. B.P.Radhakrishna, Jour.Geol.Soc.Ind.Vol.8, p.102-109 (1967).
11. J.Swami Nath et al, Int.Seminar, Tectonics and Metallogeny of SE Asia and Far East Abstracts, p.38-39 (1974).
12. B.P.Radhakrishna, Jour.Geol.Soc.India,Vol.15, p.439-454 (1974).
13. J.Swami Nath, M.Ramakrishnan and M.N.Viswanatha, Rec.Geol.Surv. India Vol.107 pt.2, p.149-175 (1976).

14. S.V.P.Iyengar, GSI Misc.Publication No.23, Precambrian Geology of the Peninsular Shield pt.2, p.415-456 (1976).
15. S.M.Naqvi and S.M.Hussain, Chem.Geol.Vol.10, p.109-135 (1972).
16. S.M.Naqvi.,V.Divakara Rao and Harinarain,Precambrian Research Vol.1, p.345-398 (1974).
17. V.S.Venkatasubramanian, Jour.Geol.Soc.India.Vol.15, p.463-468 (1974).
18. C.S.Pichamuthu, Jour.Geol.Soc.India.Vol.15(1), p.1-30 (1974).
19. A.S.Ramiengar, M.Ramakrishnan and M.N.Viswanatha and V.Srinivasa Murthy, Jour.Geol.Soc.Ind.Vol.19.No.12, p.531-549 (1978).
20. Brian Chandwick, M.Ramakrishnan, M.N.Viswanatha and V.Srinivasa Murthy,Jour.Geol.Soc.Ind.Vol.19,No.12, p.531-549 (1978).
21. Geology and Mineral Resources of the States of India, GSI Miscellaneous Publication No.30.
22. M.Ziauddian and S.Narayanaswami, Bull.Geol.Surv.India Series-A-Economic Geology No.38 pt.I. (1974).
23. S.Narayanaswami, Proc.of the Second Symposium on Upper Mantle Project, (1970).

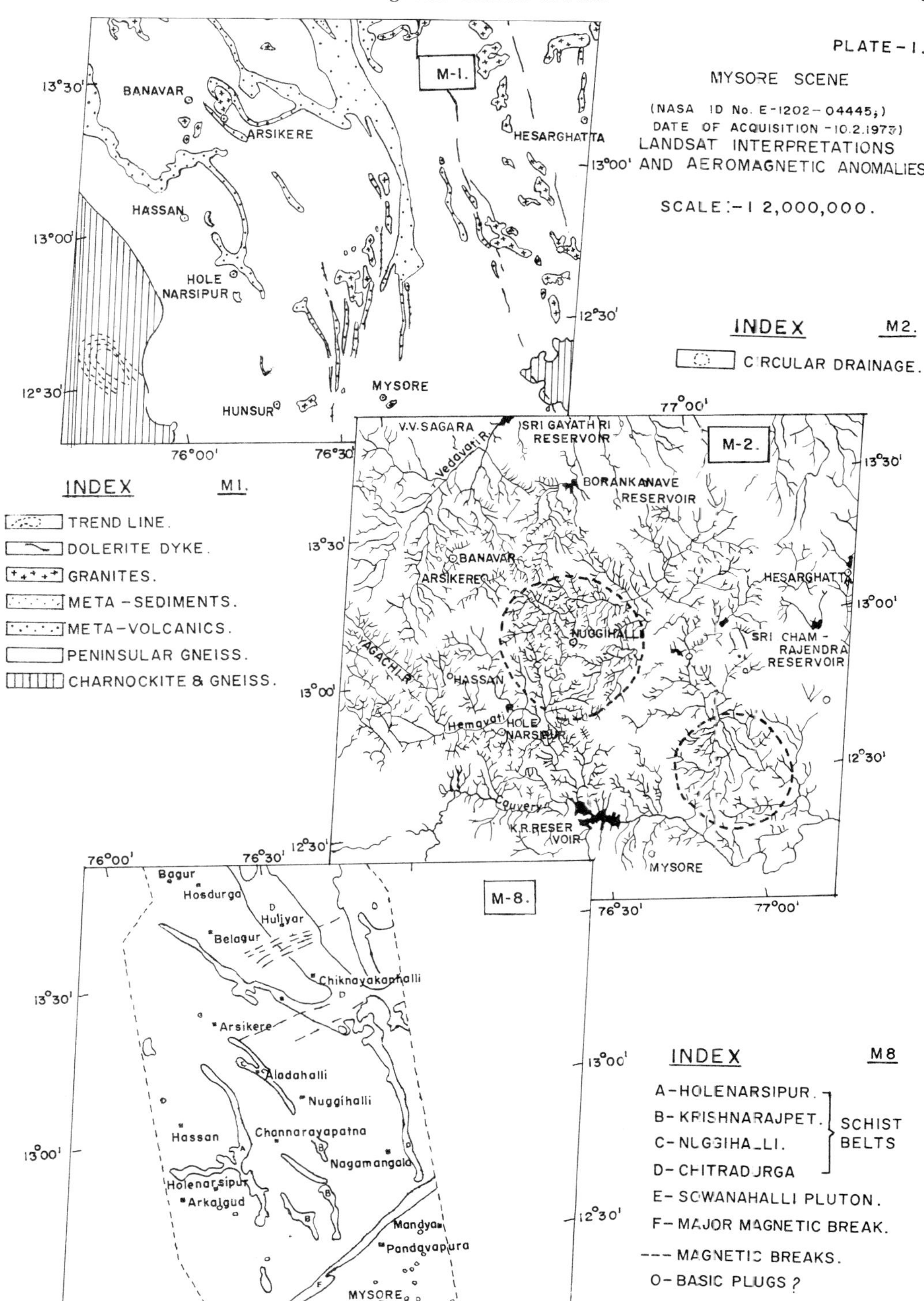
PLATE - I.
MYSORE SCENE
(NASA ID No. E-1202-04445,)
DATE OF ACQUISITION - 10.2.1973)
LANDSAT INTERPRETATIONS AND AEROMAGNETIC ANOMALIES
SCALE :- 1 2,000,000.
M-1.
BANAVAR
ARSIKERE
HASSAN
HOLE NARSIPUR
HUNSUR
MYSORE
HESARGHATTA
13°30'
13°00'
12°30'
76°00'
76°30'
INDEX M1.
TREND LINE.
DOLERITE DYKE.
GRANITES.
META - SEDIMENTS.
META - VOLCANICS.
PENINSULAR GNEISS.
CHARNOCKITE & GNEISS.
INDEX M2.
CIRCULAR DRAINAGE.
M-2.
77°00'
V.V.SAGARA
Vedavati R.
SRI GAYATHRI RESERVOIR
BORANKANAVE RESERVOIR
BANAVAR
ARSIKERE
HESARGHATTA
NUGGIHALLI
SRI CHAMRAJENDRA RESERVOIR
YAGACHI R.
HASSAN
Hemavati
HOLE NARSIPUR
Cauvery R.
K.R.RESERVOIR
MYSORE
M-8.
Bagur
Hosdurga
Huliyar
Belagur
Chiknayakanhalli
Arsikere
Aladahalli
Nuggihalli
Hassan
Channarayapatna
Nagamangala
Holenarsipur
Arkalgud
Mandya
Pandavapura
MYSORE
INDEX M8
A - HOLENARSIPUR.
B - KRISHNARAJPET.
C - NUGGIHALLI.
D - CHITRADURGA
SCHIST BELTS
E - SOWANAHALLI PLUTON.
F - MAJOR MAGNETIC BREAK.
--- MAGNETIC BREAKS.
O - BASIC PLUGS ?

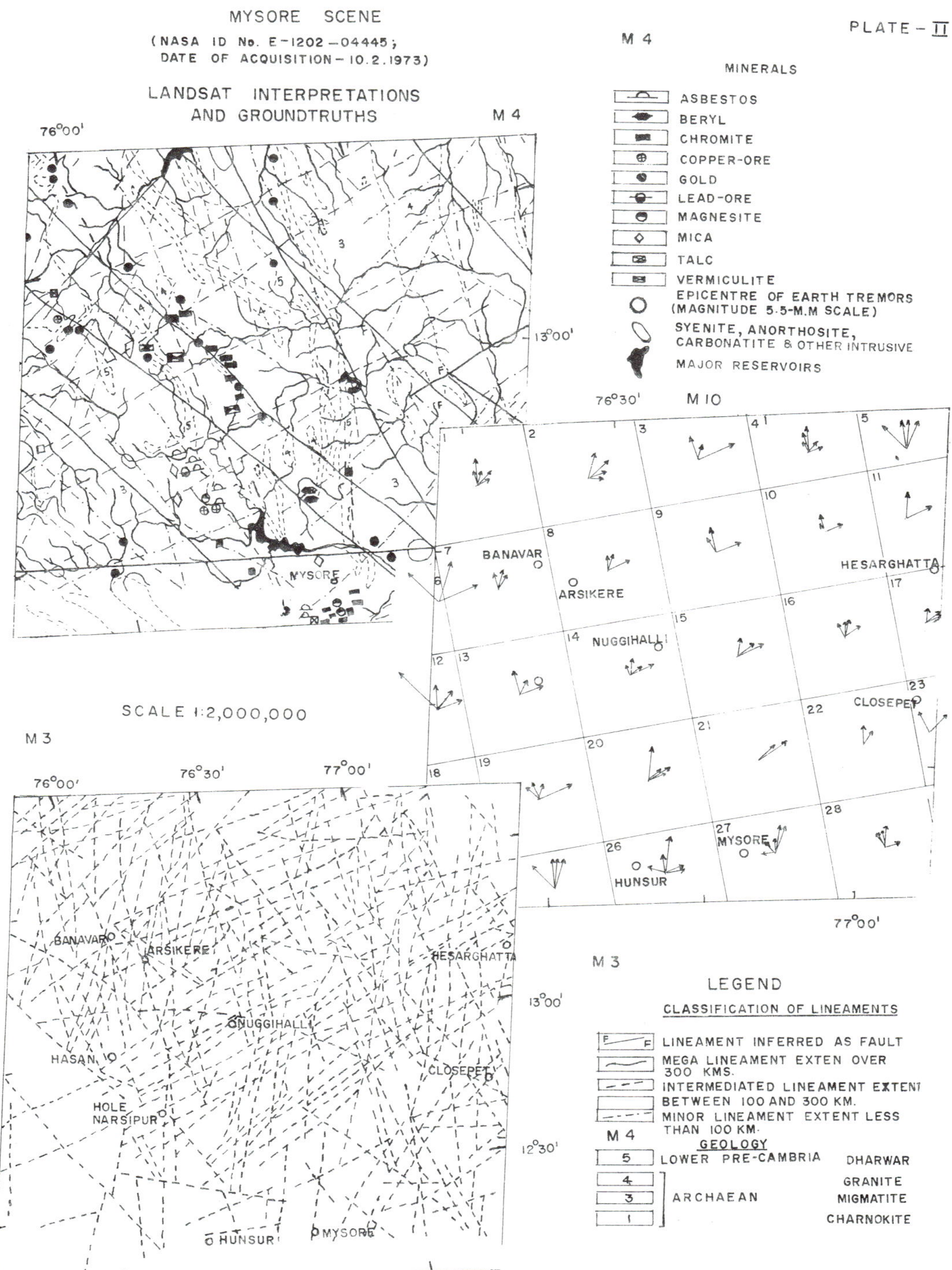
MYSORE SCENE
(NASA ID No. E-1202-04445;
DATE OF ACQUISITION-10.2.1973)
LANDSAT INTERPRETATIONS
AND GROUNDTRUTHS
PLATE-II
M 4
76°00'
13°00'
MINERALS
ASBESTOS
BERYL
CHROMITE
COPPER-ORE
GOLD
LEAD-ORE
MAGNESITE
MICA
TALC
VERMICULITE
EPICENTRE OF EARTH TREMORS
(MAGNITUDE 5.5-M.M SCALE)
SYENITE, ANORTHOSITE,
CARBONATITE & OTHER INTRUSIVE
MAJOR RESERVOIRS
MYSORE
76°30'
M 10
BANAVAR
ARSIKERE
HESARGHATTA
NUGGIHALLI
CLOSEPET
MYSORE
HUNSUR
77°00'
SCALE 1:2,000,000
M 3
76°00'
76°30'
77°00'
BANAVAR
ARSIKERE
HESARGHATTA
NUGGIHALLI
HASAN
CLOSEPET
HOLE
NARSIPUR
HUNSUR
MYSORE
13°00'
12°30'
LEGEND
CLASSIFICATION OF LINEAMENTS
LINEAMENT INFERRED AS FAULT
MEGA LINEAMENT EXTEN OVER
300 KMS.
INTERMEDIATED LINEAMENT EXTENT
BETWEEN 100 AND 300 KM.
MINOR LINEAMENT EXTENT LESS
THAN 100 KM.
GEOLOGY
LOWER PRE-CAMBRIA
DHARWAR
ARCHAEAN
GRANITE
MIGMATITE
CHARNOKITE

LANDSAT EXPLORATION OF HIMALAYAN AND PENINSULAR REGIONS (REMOTE SENSING AND MINERAL EXPLORATION — PROGRESS REPORT OF WORK DONE IN INDIA) — I.G.C.P. PROJECT 143

Presented by B. N. Raina, National Convener

Based on the work carried out by H. N. Shukla, S. B. Sharma, A. K. Roy, A. V. Raju, M. Ahmed, P. Rao, V. D. Bhate, D. S. Kamet et al. and K. S. Dasgupta et al.

ABSTRACT

Work on this project was started in India in August, 1977. Six project areas were selected for such studies in areas of known mineralization. Five of these projects are located on the Precambrian rocks of the Indian Peninsular Shield and one in the Himalayan area. Work carried out so far includes collection of regional and detailed geological data, interpretation of regional structure and lineaments from LANDSAT images and preparation of combined maps on which an attempt has been made to integrate the LANDSAT data with the conventional data. In most cases it has been possible to establish relationships between the reported mineral occurrences and areas of intersection between major lineaments and/or curvilinear features. In some cases such studies have thrown new light on the regional structure of the area and possible genetic relationship between the mineral occurrences, structure and lithology of the area.

INTRODUCTION

As Convener of the I.G.C.P. Project 143 in India, a review of the work done in India on this Project is being presented. Six project areas were selected for study, one of them lies in Himalaya and the rest of them are in Indian Peninsula. Interpretation of a few frames of LANDSAT imagery, which help in elucidating the regional structure of the Himalaya, have also been included in this project report. Work on this project was started in India in August, 1977. The present paper gives the progress of work done on some of these test sites, and our tentative conclusions. The work was carried out by my colleagues in the Geological Survey of India, who were assigned this job in addition to their normal duties.

METHODOLOGY

The LANDSAT images of the test areas in four bands - 4, 5, 6 and 7 - have been interpreted individually with the help of a hand lens in both transmitted as well as reflected light. The images have also been examined under mirror stereoscope in pairs of Bands 4 and 5 and in Bands 6 and 7. In a few cases some amount of spatial filtering (visual) has also been obtained by the use of Ronchi Ruling. The interpretations made on individual frames have been transferred on to a geological base map of the area on 1:1,000,000 scale to make a composite map. On this locations of known base metal occurrences have been plotted to obtain an integrated picture of the geology, photo-linears and mineral occurrences of the test sites. In some cases, where the test area is of limited extent, scale of the map given is 1:250,000.

LINEAMENT CLASSIFICATION

While classifying the lineaments the workers have taken into consideration both physical and genetic parameters. On the basis of physical appearance, lineaments have been divided into (i) curvilinears - which could represent various fold closures and/or erosional/drainage characters; (ii) linears - which could represent fracture zones, fold axes and formational boundaries. Some linear features which do not fall into the above category, yet are well defined on the images, perhaps represent local faults or some basement features. These linears have further been classified as major, intermediate and minor depending on their length of 300 kms or more, 100 km to 300 kms and less than 100 kms respectively.

RESULTS OF STUDY ON DIFFERENT PROJECTS

Mineralized Belts in Himalaya

So far very few economically viable base metal deposits have been located from the Himalaya. The reason for this enigma is not quite clear. With all the parameters present, including structural controls, Precambrian meta-sediments, vast stretches of carbonates, volcanics and magmatic activity etc., it becomes difficult to resign oneself to the idea that Himalaya is bereft of base metal mineralization. Hence renewed efforts are always being made to test fresh ideas in this virgin and frustrating terrain. Two test sites were, therefore, selected in the Himalaya for study under this Project - one lies in the Lesser Himalayan zone of U.P. and the other in the zone of proposed 'Himalayan Suture' in Kashmir.

Mineralized Belt of U.P. Lesser Himalaya

Dhanpur-Pokhri and adjacent areas of U.P. Field work done on these occurrences has indicated that the copper, lead, antimony, zinc are associated with both autochthonous (Askot) litho-masses. These mineral occurrences are of very limited strike extension (less than 1,000 m.). Monocular visual interpretation of the LANDSAT 1 imagery of the region was carried out in bands 4, 5, 6 and 7. This has brought to light hitherto undetected linear and curvilinear features. The linear features on the basis of their regional trend and probable order of antiquity can be grouped as follows: (i) E-W (ii) NW-SE to WNW-ESE (iii) NE-SW to N 10°E/N 10°W - S 10°W/S 10°E. The latest transverse set of lineaments with respect to the regional structural grain of Himalaya, have affected even the Neogene-Quarternary sequence of Siwalik. The Lesser Himalaya represent a zone of intense tectonic activity as seen by the numerous and haphazard arrangement of linears and curvilinears on the LANDSAT data. An integrated map of the area showing major lineament, structural elements, broad lithological groups and mineral occurrences has been prepared (Fig. 1). From this map, it is seen that the mineral occurrence of autochthonous belt lie close to the Main Central Thrust and may be genetically related to it. As mineralization in the area seems to be structurally controlled, locales, where more than two sets of curvilinear/linear features intersect, represent potential areas for base metal exploration.

Indus suture and adjoining areas. A highly tectonized belt of rocks, ranging in age from Late Cretaceous to Mid-Miocene occurs along the Dras-Indus Valley in Kashmir on a Pre-Cambrian (?) granitic basement. The rocks are ophiolites, limestones, rhythmites, conglomerates, flysch and molasse. Being a tectonized zone and possibly a plate junction, it is considered to have great potential from economic mineral point of view. Although, reports of occurrence of gold, chromite and copper from the area exist in earlier records, investigations so far have revealed only a very small deposit of chromite. Interpretation from the four bands of LANDSAT imagery reveals a number of linears and curvilinears in the area (Fig. 2). On the basis of various sets of lineaments delineated, these can be grouped as per their probable order of antiquity as follows:
(i) NW-SE (ii) N 10°W - S 10°E to N 10°E - S 10°W and (iii) NE - SW to ENE-WSW. These linears are particularly abundant in the Pre-cambrian (?). It is also believed that the granite of the area are not entirely Precambrian granitic basement but many later phases, including if Tertiary period are present. The area has great potential for porphyry copper and the major NW-SE trending lineaments may prove significant in this context.

Mineralized Belts of Indian Peninsular/Shield Area

Major base metal occurrences are confined to the Indian Shield area where economic deposits of lead, copper, zinc, gold, etc. are found. Hence, most of the test sites have been located in the shield area.

Rajpura-Dariba belt. Old workings and presence of ferruginous breccia in Rajpura-Dariba has been recorded earlier. The mineralization in Rajpura-Dariba area occurs in the high grade metamorphic rocks of Pre-Aravalli age (Proterozoic?) which form low north-south longitudinal ridges. The general structural trend is N-S to N 10°E - S 10°W in the Dariba area, changing to N 20°E - S 20°W northwards and then to N 50°E - S 50°W in Bethumni area. Dips vary from 60° to 80° towards east and southeast. Fold closures show plunges of 30° - 60° towards north-east. Faults parallel to bedding and dipping 65° - 80° towards east are present in the area. Mineralization in the area is considered to be chemo-genic type syngenetic with the deposition of sediments in euxinic environments in isolated basins.

Zawar belt. The Zawar Pb-Zn mineralization occurs in Aravalli rocks of low grade metamorphites of green schist facies and the ore is found in sandy dolomites, dolomites and marly slates. The regional trend of the rocks varies from NNW-SSE to NW-SE, with local variations. The total extent of the mineralized belt is about 20 kms. The mineralization is of two types: (i) sedimentary bedded type and (ii) metamorphic remobilized type. The examination of LANDSAT imagery of the region, with some enhancement obtained by Ronchi Ruling, shows the dominant trend of lineaments in the mineralized areas (Fig. 3) to be ENE-WSW with two conjugate sets of NE-SW trends, with intersections in the major and minor groups. Some important NNW-SSE trending sets have also been marked in the minor group of lineaments. The lineaments are usually cutting across lithological as well as stratigraphic boundaries. Since a syn-sedimentary chemogenic origin has been suggested for the mineralization, it is not clear whether any particular relationship exists between the lineaments and the mineralization. However, as the rock types occur in strongly tectonized linear belts with a regional NE-SW trend, it is possible that sedimentation and folding of the rocks as well as the movement of hydrothermal fluids may have been controlled by the dominant ENE-WSW and NNW-SSE lineaments.

Mailaram copper belt. This mineralized belt occurs in the Khammam district of Andhra Pradesh. The important occurrences are, Mailaram Yellambailu, Panjar and Sarakal. Mineralization is found in the Pre-Cambrian rocks of Dharwar Group and overlying metamorphites of Pakhal Group of rocks. The most important mineralization is associated with grey quartz-veins traversing quartz-chlorite schist of Dharwar Group. The veins are emplaced along a series of NW-SW trending dextral and sinistral *en echelon* zones of shearing and brecciation parallel to the axes of drag folds related to the major folds. The association of copper minerals with quartz veins and with sulphides, like pyrrhotite and molybdenite, their mode of occurrence and the wall rock alteration strongly point to the hydrothermal origin. The mineralization show structural and lithological controls.

Interpretation of LANDSAT imagery shows four major lineaments and a number of minor lineaments in the area (Fig. 4). Three of these major lineaments have a NW-SE trend, while the fourth one has a NW-SW trend, which was also identified as the Mallavaram fault from the field work. The strike of the mineralized shear zones of Mailaram area, the Mallavaram fault and the axes of the major folds in the area show parallelism, indicating possible relationship to the same tectonic episode. It is also interesting to record that the copper occurrences of Mailaram and Yellambailu lie more or less along one of the NW-SE trending lineaments, whereas the Banjar and Sarakal "occurrences are parallel to the NE-SW lineaments."

Gani-Kalva copper belt. It is an E-W trending narrow belt, 27 kms long and 48 kms. wide in the Cuddapah district of Andhra Pradesh. The mineralization is found in the shales and associated basic sills of Tadpatri Formation which forms part of the Pre-Cambrian Cheyyar Group of Cuddapah Super Group. The regional trend of the rocks varies from E-W to ENE-WSW. The Gani-Kalva area represents a drag folded asymmetrical anticline, which has low plunge towards N 80°E. The axis of the fold follows the major Gani-Kalva fault which is considered to be a basement fault, reactivated from time to time. The Gani-Kalva anticline was refolded into a number of dextral and sinistral drag folds and is accompanied by a number of shear faults and tensional fractures. The overall picture seems to indicate that the mineralization is controlled by tension fractures developed alongside the shear fault. Interpretation of LANDSAT imagery (Fig. 5) has brought out 2 major lineaments in the area. One of them trends ENE-WSW and can be traced for over 250 kms., and coincides with the known Gani-Kalva Fault. The other lineament trends in NE-SW direction and can be traced for a distance of 75 kms. A number of minor lineaments are also inferred in the area, most of them showing NW-SE trends, coinciding with the master-joints in the quartzites of the area. It is highly probable that the Gani-Kalva lineament exerted considerable control on the movement of mineralized fluids in the area and, therefore, should be examined with special attention. Specially the area of intersection of this lineament with the NE-SW and NW-SE trending lineaments deserves careful investigation.

Bastar mineralized belt. This area forms part of the Archaen terrain of Madhya Pradesh. It is predominantly occupied by rocks of Archaen and Precambrian age, except for a narrow belt in the Godavari valley in the south, which contains rocks of Gondwana age. A number of pegmatitic veins traverse the oldest metasediments and associated basic rocks of the Bengpal Group. These pegmatites carry cassiterite mineralization and are considered to be related to the intrusion of Paliam and Darba granites. Minor incidence of cassiterite has also been noticed in the above granites. The tin mineralization occurs in the form of discrete crystals and thin veins of cassiterite associated with albite, lepidolite, beryl, flourite, tourmaline, columbite, tentallite and magnetite. Two distinct zones of pegmatites with a regional trend of NNW-SSE to NE-SW have been recognized in the area, extending for a length of over 15 km. each. The first set of pegmatites are zoned and contain higher incidence of lepidolite and beryl and a mixed mineralization of Sn and Ta-Nb. The second set of pegmatites exhibiting griesening carry higher incidence of cassiterite mineralization ranging from 0.3 to 1%. Detailed exploration has resulted in location of 149 pegmatites, out of which 17 are found to be cassiterite bearing. Four LANDSAT images were interpreted for the area, the results of which are shown in the integrated map (Fig. 6). Several lineaments have been picked up, which show no appreciable variation in their trends from one cover type to another. Certain areas do show greater density of the lineaments. The major lineament trends are NW-SE, NE-SW, NNE-SSW, E-W, N-S and ENE-WSW, the predominant trend being NW-SE. This is also the trend of the basic dykes and quartz reefs traversing the area. Most of these lineaments reflect the joint and fracture pattern of the area, except the major and intermediate lineaments which are considered to represent major faults and shear zones. The reported occurrence of the cassiterite bearing pegmatites in the Govindpal-Mundaval-Chiruwanda area lie close to a WNW-ESE trending lineament, which perhaps also represents a fault, offsetting some geological formations. The pegmatites showing better concentration of cassiterite near Mundaval are also aligned in the same general direction as this lineament.

Malanj Khand copper belt. The Malanj Khand copper belt of Madhya Pradesh, is also known from its old workings. The area forms part of the Archaen terrain. The rock formations exposed are granitic rocks of Basement complex, intruded by quartz reefs and basic dykes. The basement crystallines are unconformably overlain by the metasediments of Chilpi Group. The Malanj Khand hill is composed of a quartz reef in the basement crystallines. It has an arcuate shape and extends for 2.5 km. in a general N-S direction, with a maximum width of 50 metres. It appears to have been emplaced along a shear plane. The basic dykes traversing the granite and quartz-reef, have NNW-SSE trend and are metamorphosed to epidiorite and amphibolite. The mineralization is mainly confined to the Malanj Khand reef and its off-shoots with some disseminations in the granite. The main ores are chalcopyrite and pyrite constituting over 90% of the ore. Chalcocite and molybdenite also occur in small quantities. Native copper is noticed in the transition zone. Wall rock alteration is seen. The reef is intensely sheared, fractured and brecciated. The mineralization is localized in the shear and fracture planes and is considered to be of hydrothermal origin. Three major lineament groups are observed on the LANDSAT imagery of the area (Fig. 7). These are (i) the north central and north-western zone with predominant ENE-WSW trends, probably related to the Satpura trend (ii) the eastern and south-eastern zone with dominant linear trends in NW-SE and NNW-SSE direction reflecting the Mahanadi-Godavari trend and (iii) the south-central and south-western zone with ENE-WSW to NE-SW, N-S to NNE-SSW and NW-SE trends. A NNE-SSW trending lineament is traceable from near Malanj Khand to Nandora in the south. This lineament controls part of the middle course of the Son river. The copper deposit of Malanj Khand and other minor occurrences of copper further south lie close to this lineament may possibly be related to it. It is likely that this lineament may be a branch of a N-S trending fault - the Kotri Darekasa fault - which represents a major curvilinear lineament zone of the area.

Panna diamond belt. Panna area is well known in India as the only diamond productive belt of the country. This diamond belt extends over a length of 90 kms. with varying width of 15 to 50 kms. within the Vindhyan Group of rocks. The primary source of the diamond in this belt are the kimberlite pipes of Majhgawan and Hinota, which are intrusive into the Kaimur sandstone, forming the lower part of the Upper Vindhyan Group. The Majhgawan pipe is pear shaped in outline and measures 500 x 830 metres. The Hinota pipe is near circular in shape and measures 215 x 180 metres. The pipe rocks are similar in composition comprising massive basaltic and micaceous kimberlite grading to autolithic kimberlite breccia, which becomes predominant at depth. Diamonds are also found as placers in the conglomerate horizons of the Vindhyan Super Group. Examination of the LANDSAT imagery of the area (Fig. 8) shows the lineaments developed in the major cover types vary in their orientation and magnitude. The lineaments over the Bundelkhand granite massif trend mainly NNE-SSW to NE-SW and NW-SE. The dominant trends in the Vindhyan sedimentaries are ENE-WSW, NE-SW, NNW-SSE and NW-SE. The Sidhi group of crystallines of the Son valley show lineaments prominantly in ENE-WSW direction. On the Deccan volcanics the lineament trends are ENE-WSW, NE-SW and NW-SE. The most striking feature is the predominance of the ENE-WSW trending lineaments in the major part of the area. In the Vindhyan sediments two interesting major lineaments have been observed near the northern margin. One of these lineaments trends ENE-WSW, passing through Panna town, while the other trends NE-SW and runs south of Panna. These two lineaments intersect west of Majhgawan Railway station, where another lineament coming from west of Rewa and trending NW-SE also intersects these lineaments. The area of intersection of these three lineaments merits detailed investigation. The kimberlite pipe of Majhgawan falls on the westerly extension of the ENE-WSW trending lineament which passes through Panna. It is worthwhile to explore the possibility of bearing of this lineament on the emplacement of the kimberlite pipe in the area. Two major lineaments observed in the Narmada-Son valley have NE-SW and ENE-WSW trends. These two lineaments together with other parallel to subparallel minor lineaments appear to constitute the Narmada-Son rift system. It is also interesting to record that a number of base metal occurrences reported from the area are confined between the two major lineaments and are in all probability genetically related to them.

INTERPRETATION OF REGIONAL GEOLOGY AND STRUCTURE OF HIMALAYA - SOME EXAMPLES

The author carried out visual interpretation of six LANDSAT images of Himalaya to verify some of the theories about the structure and tectonics of this range. The areas covered by these images lie in parts of Kashmir, Himachal Pradesh and Arunachal states of India. One set of images covers the area of the western syntaxial bend in the Muzaffarabad-Nanga Parbat area of Kashmir and shows the hair-pin loop (Fig. 9) taken by the rocks and the Murree and Panjal thrusts between the Kunihar river and Muzaffarabad. The gradual increase in the angle of convergence of fold axes to the north and south of this area is clearly visible. The final disappearance of this effect is seen in the south, near Kotli. The synoptic view given by these images clearly shows this tectonic situation and thus confirms Wadia's hypothesis of syntaxial bend in the area. Fig. 10 shows the eastern extremity of the Himalaya lying in Arunachal Pradesh. The images show a wide arcuate bend of the ranges from a trend of E-W to NW-SE and finally to N-S in the mountaineous catchment of the Brahmaputra river. The siwaliks in the foot hill zone also take a turn from Pasighat on the Dihang to Nizamaghat on the Dihang and finally disappear below the crystalline cover. Due to lack of detailed geological knowledge of the area, it is difficult presently to confirm whether a syntaxial type of bend occurs in the area or not. However, from the images it is seen, that an orographic arcuate bend is present here.

MACHINE PROCESSING OF THE DATA

Active collaboration was sought from ISRO for development of hard- and soft-ware for machine processing of the LANDSAT data, as a follow-up action after the initial visual interpretation of the LANDSAT imagery has been carried out. The ISRO submitted two technical notes. In the one they have developed a set of computer programmes for carrying out ratioing techniques to improve the definition of geological features. Significant contrasts have been obtained by using ratios of Band 4 to Band 7 and Band 5 to Band 7. In the future programme, ratioing of Band 4 to Band 5, Band 4 to Band 6 and generation of colour composites is to be carried out to study the project areas with the help of these enhancement techniques. The other technical note deals with the man-machine interactive system for extraction of linears. The hardware and software of this system was designed and constructed at the Space Application Centre, Ahmedabad. With the help of a Video Display and Graf/Pen devices and using this with a TPA-70 computer it is possible to digitize and obtain a graphical display of geological features like lineaments. The digitized data on the photo linears can also be processed to determine the statistical distribution of linears.

CONCLUSIONS AND RECOMMENDATIONS

The present study involved only the visual interpretation of the LANDSAT imagery in the four spectral bands - No. 4 (0.5-0.6μ), No. 5 (0.6-0.7μ), No. 6 (0.7-0.8μ) and No. 7 (0.8-1.1μ) on an approximate scale of 1:1,000,000 in most cases. The imagery is in the form of black and white contact paper prints with format size - 24 cm. x 24 cm. (9.5" x 9.5"). Each frame on the million scale represents a coverage of approximately 185 km. x 185 km. The imagery was obtained in the form of black and white negatives from the ISRO and their contact prints were reproduced for study. The successive reproduction from the original has considerably degraded the quality of the imagery. The deterioration varies from band to band but is more pronounced in bands 4 & 5. Mostly band 7 of the infra-red region has been found to be most useful for geological interpretation, though in some cases band 5 has also given good contrasts. In most cases it has been possible to interpret broad geological formations, igneous rocks and major structural details on these images by their tonal contrasts, texture, geomorphological expressions and drainage patterns. The principle applied for exploration of mineral by remote sensing technique is to use the conventional parameters of mineral exploration as the basis of interpreting the remote sensing data. The more important guides or parameters are the geometry or symmetry of geological features, apparent geological anomalies and regional trends. Other factors affecting mineral exploration are relevant minerological or geochemical patterns and anomalies of vegetation, soil and drainage patterns. The coincidental occurrence of two or more favorable factors such as faulting or intrusion meeting receptive lithology or structure often indicates a favorable location for exploration. Although some mineral deposits are related to fractures, most fracturing is unrelated to mineralization. Nevertheless, identification of previously unknown lineaments and fracture systems is the biggest contribution of LANDSAT imagery in mineral exploration. In the case histories outlined above it is seen that in some cases a probable correlation between the lineaments or fracture patterns and the known mineral occurrences may exist. Thus there is a strong probability of mineralization seen in the Deoban/ Garhwal rocks of Dhanpur-Pokhri area with the prominent joint and shear planes observed on the LANDSAT imagery. The copper occurrence of the Mailaram area may be genetically connected to the NNE-SSW trending Mailaram Fault of the area, as the mineralization is considered to be of hydrothermal origin. Similarly the mineralization of Gani-Kalva area, which is mainly controlled by tensional fractures developed in the axial region of the anticline, may be related to the main ENE-WSW trending Gani-Kalva fault. At least the development of the tensional fractures is thought to be the result of this fault. In Panna area one of the major lineament passes through the Panna town, close to the kimberlite pipe area, and may have some bearing on the emplacement of the pipe. In Bastar area pegmatites showing better concentration of cassiterite, near Mundaval, are aligned in the general direction of a major lineament. The best example is afforded by the Singhbhum copper belt (Fig. 11), where the productive copper occurrence are dotted along a NW-SE trending major lineament. The detailed work by Geological Survey of India has shown that here the mineralization follows the shear zones irrespective of the country rocks. It is also found that the mineralization is generally accompanied by hydrothermal alteration. The lodes are generally lens shaped and plunge parallel to the fold axes in a general NNE to NE direction and most of the linears in the area plunge at 35° to 50° toward NNE to NE. The structural systems defined mainly by the trends of linears, their size and spacing, seem to be directly related to the physical properties of the lithosphere and the deformational evolution of the region. The question of relationship of mineral deposits and the circulation of magmatic mineralizing fluids along geocrustal fractures has also received much attention. Emphasis has also been placed on the role of the intersection of regional lineaments in the emplacement of mineral deposits. This together with the concept of lithological control and of selective mineralization of host rock, which may be registered on space imagery as linear or circular tonal anomalies, serve as useful guides for mineral exploration by remote sensing techniques. Due to various reasons the work carried out so far was limited to collection of existing data on the project sites and visual interpretation of the LANDSAT imagery. In the future programme for this project the following studies have been envisaged:

1. Computer enhancement of tapes and ratio image techniques in limited areas.
2. Colour-enhanced band ratio techniques, if possible.
3. Colour-composite studies with additive colour viewer with 70 mm chips for selective areas for more detailed analysis.
4. Principal component analysis to achieve enhancement of structural features.
5. Linear extraction studies with the help of Man-Machine interaction system developed by ISRO.

ACKNOWLEDGEMENTS

The author is thankful to N.A.S.A., U.S.A.; I.S.R.O., Ahmedabad; I.I.T., Bombay; I.I.T., Kharagpur; O.N.G.C., Dehradun; N.R.S.A., Hyderabad; N.G.R.I., Hyderabad; A.M.S.E. Wing, G.S.I. and Regional Integrated Survey Divisions of G.S.I. and the different coordinators of the projects for their help and cooperation received in the conducting of this project and preparation of the reports. Due to shortage of space, details of large number of publications consulted are not included in this report. Nevertheless their help and guidance in preparation of this report is gratefully acknowledged.

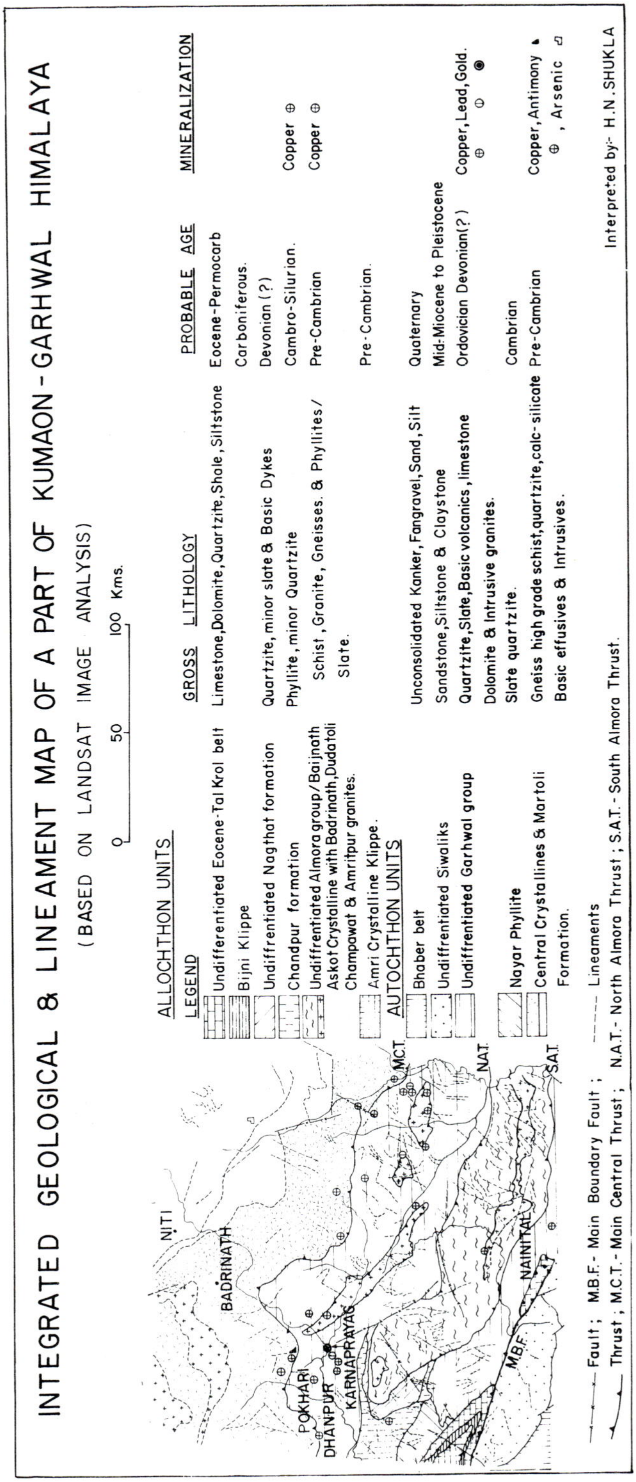

INTEGRATED GEOLOGICAL & LINEAMENT MAP OF A PART OF KUMAON-GARHWAL HIMALAYA

(BASED ON LANDSAT IMAGE ANALYSIS)

0 50 100 Kms.

NITI
BADRINATH
POKHARI
DHANPUR
KARNAPRAYAG
MCT
NAT
M.B.F.
NAINITAL
SAT

ALLOCHTHON UNITS

LEGEND	GROSS LITHOLOGY	PROBABLE AGE	MINERALIZATION
Undifferentiated Eocene-Tal Krol belt	Limestone, Dolomite, Quartzite, Shale, Siltstone	Eocene-Permocarb	
Bijni Klippe		Carboniferous.	
Undiffrentiated Nagthat formation	Quartzite, minor slate & Basic Dykes	Devonian (?)	
Chandpur formation	Phyllite, minor Quartzite	Cambro-Silurian.	Copper ⊕
Undiffrentiated Almora group/Baijnath Askot Crystalline with Badrinath, Dudatoli Champawat & Amritpur granites.	Schist, Granite, Gneisses. & Phyllites/ Slate.	Pre-Cambrian	Copper ⊕
Amri Crystalline Klippe.		Pre-Cambrian.	
AUTOCHTHON UNITS			
Bhaber belt	Unconsolidated Kanker, Fangravel, Sand, Silt	Quaternary	
Undiffrentiated Siwaliks	Sandstone, Siltstone & Claystone	Mid-Miocene to Pleistocene	
Undiffrentiated Garhwal group	Quartzite, Slate, Basic volcanics, limestone Dolomite & Intrusive granites.	Ordovician Devonian(?)	Copper, Lead, Gold. ⊕ ⦶ ◉
Nayar Phyllite	Slate quartzite.	Cambrian	
Central Crystallines & Martoli Formation.	Gneiss high grade schist, quartzite, calc-silicate Basic effusives & Intrusives.	Pre-Cambrian	Copper, Antimony ▲ ⊕, Arsenic ◺

Fault; M.B.F.- Main Boundary Fault; ----- Lineaments

Thrust; M.C.T.- Main Central Thrust; N.A.T.- North Almora Thrust; S.A.T.- South Almora Thrust.

Interpreted by:- H.N.SHUKLA

FIGURE 1.

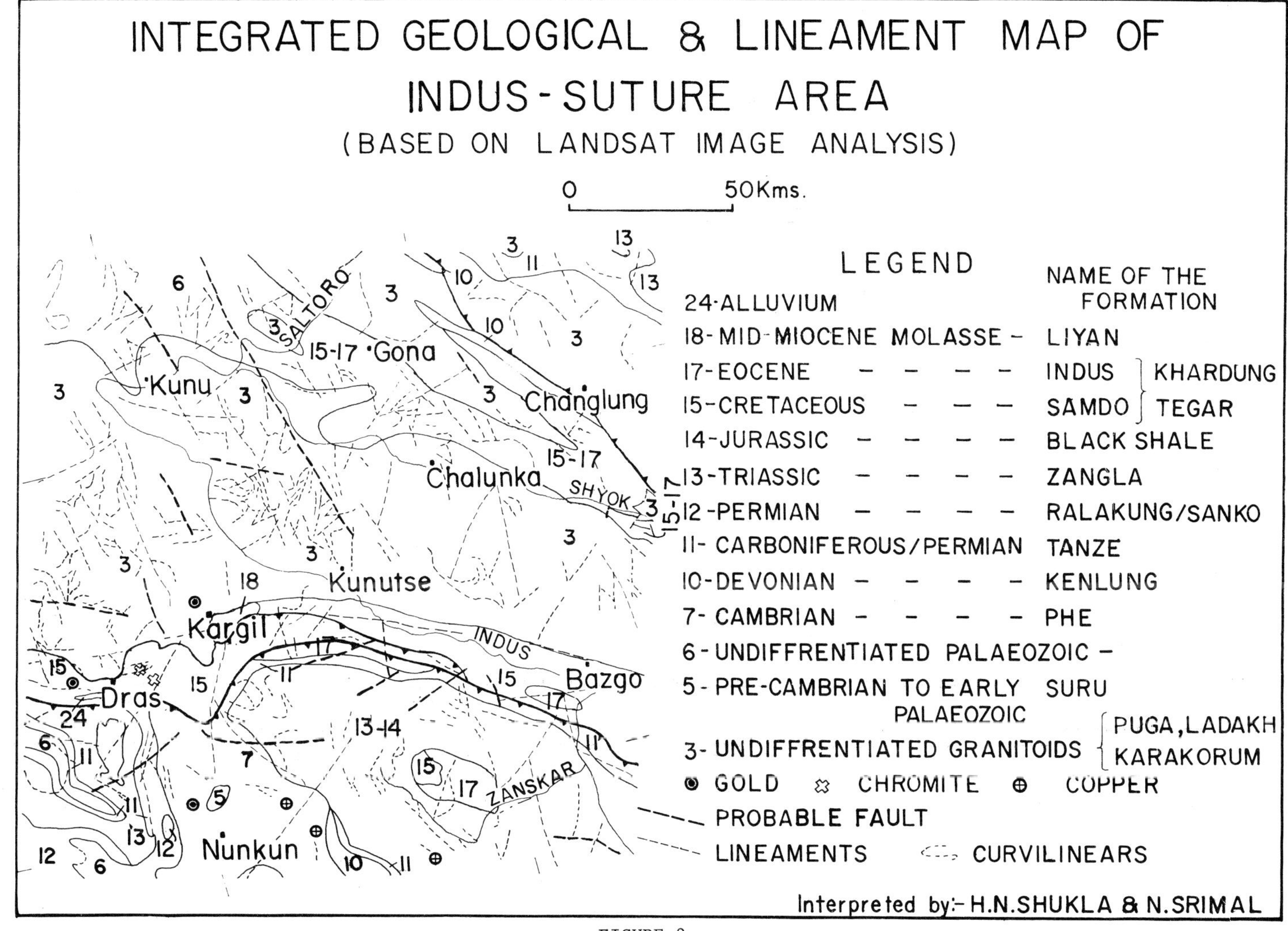
INTEGRATED GEOLOGICAL & LINEAMENT MAP OF
INDUS - SUTURE AREA
(BASED ON LANDSAT IMAGE ANALYSIS)
0 50Kms.
LEGEND
NAME OF THE FORMATION
24-ALLUVIUM
18-MID-MIOCENE MOLASSE - LIYAN
17-EOCENE - INDUS } KHARDUNG
15-CRETACEOUS - SAMDO } TEGAR
14-JURASSIC - BLACK SHALE
13-TRIASSIC - ZANGLA
12-PERMIAN - RALAKUNG/SANKO
11- CARBONIFEROUS/PERMIAN TANZE
10-DEVONIAN - KENLUNG
7- CAMBRIAN - PHE
6-UNDIFFRENTIATED PALAEOZOIC -
5-PRE-CAMBRIAN TO EARLY PALAEOZOIC SURU
3- UNDIFFRENTIATED GRANITOIDS { PUGA, LADAKH KARAKORUM
GOLD CHROMITE COPPER
PROBABLE FAULT
LINEAMENTS CURVILINEARS
Interpreted by:- H.N.SHUKLA & N.SRIMAL
SALTORO
Kunu
Gona
Changlung
Chalunka
SHYOK
Kunutse
Kargil
INDUS
Bazgo
ZANSKAR
Dras
Nunkun

FIGURE 2.

INTEGRATED GEOLOGICAL & LINEAMENT MAP OF ZAWAR & RAJPURA-DARIBA

(BASED ON LANDSAT IMAGERY INTERPRETATION)

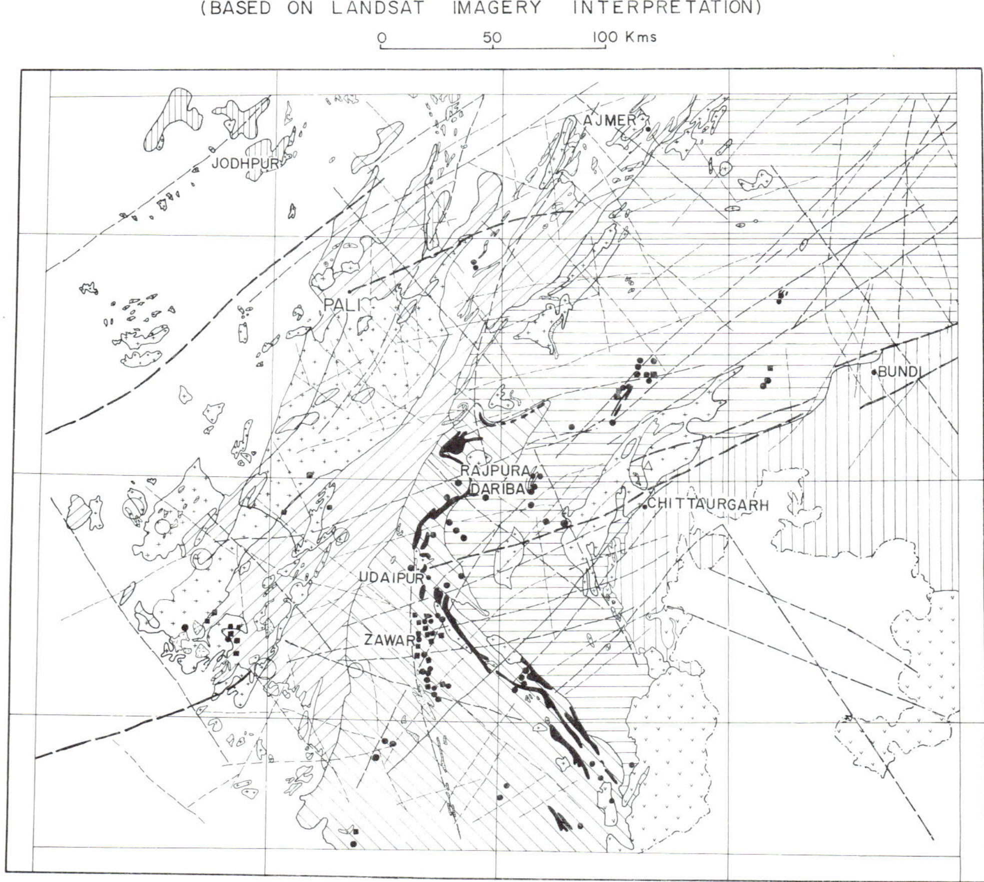

INDEX

	GROSS LITHOLOGY	AGE
ALLUVIUM / BLOWN SAND		
VINDHYAN SUPER GROUP	SANDSTONE, SHALE & LIMESTONE CONGLOMERATE	EARLY PALEOZOIC TO LATE PROTOROZOIC
DELHI SUPER GROUP	QUARTZITE SCHISTS, GNEISSES, AMPHIBOLITES MARBLE, CONGLOMERATE.	PROTOROZOIC
ARAVALLI SUPER GROUP WITH MARBLE & DOLOMITE	QUARTZITES, SCHISTS, MARBLE CONGLOMERATE	PROTOROZOIC
PRE-ARAVALLI WITH MARBLE & DOLOMITE	SCHISTS, GRANITE, GNEISSES AMPHIBOLITES QUARTZITE, CONGLOMERATE, DOLOMITE, MARBLE	ARCHAEAN

INTRUSIVES & EFFUSIVES

- DECCAN TRAP
- MALANI RHYOLITE
- GRANITES & PEGMATITES
- ULTRABASICS & BASICS

LINEAMENTS

- MAJOR
- INTERMEDIATE
- MINOR

- △ HOT SPRING
- ■ LEAD & ZINC ORE
- ● COPPER ORE
- ○ EPICENTRE MAGNITUDE <5
- ◯ EPICENTRE MAGNITUDE >5

FIGURE 3.

INTEGRATED LINEAMENT & GEOLOGICAL MAP OF MAILARAM COPPER BELT

(BASED ON INTERPRETATION OF LANDSAT IMAGERY)

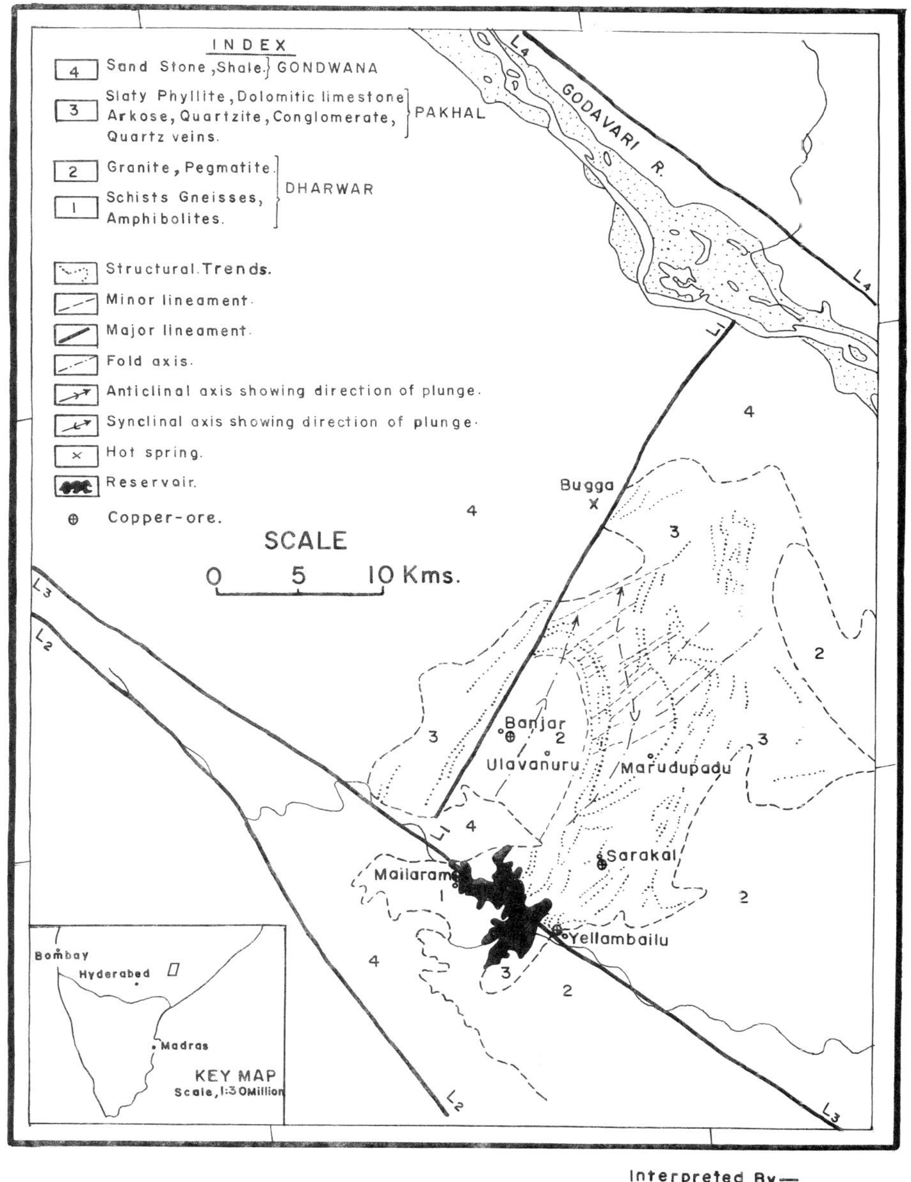

Interpreted By—
A.V. RAJU

FIGURE 4.

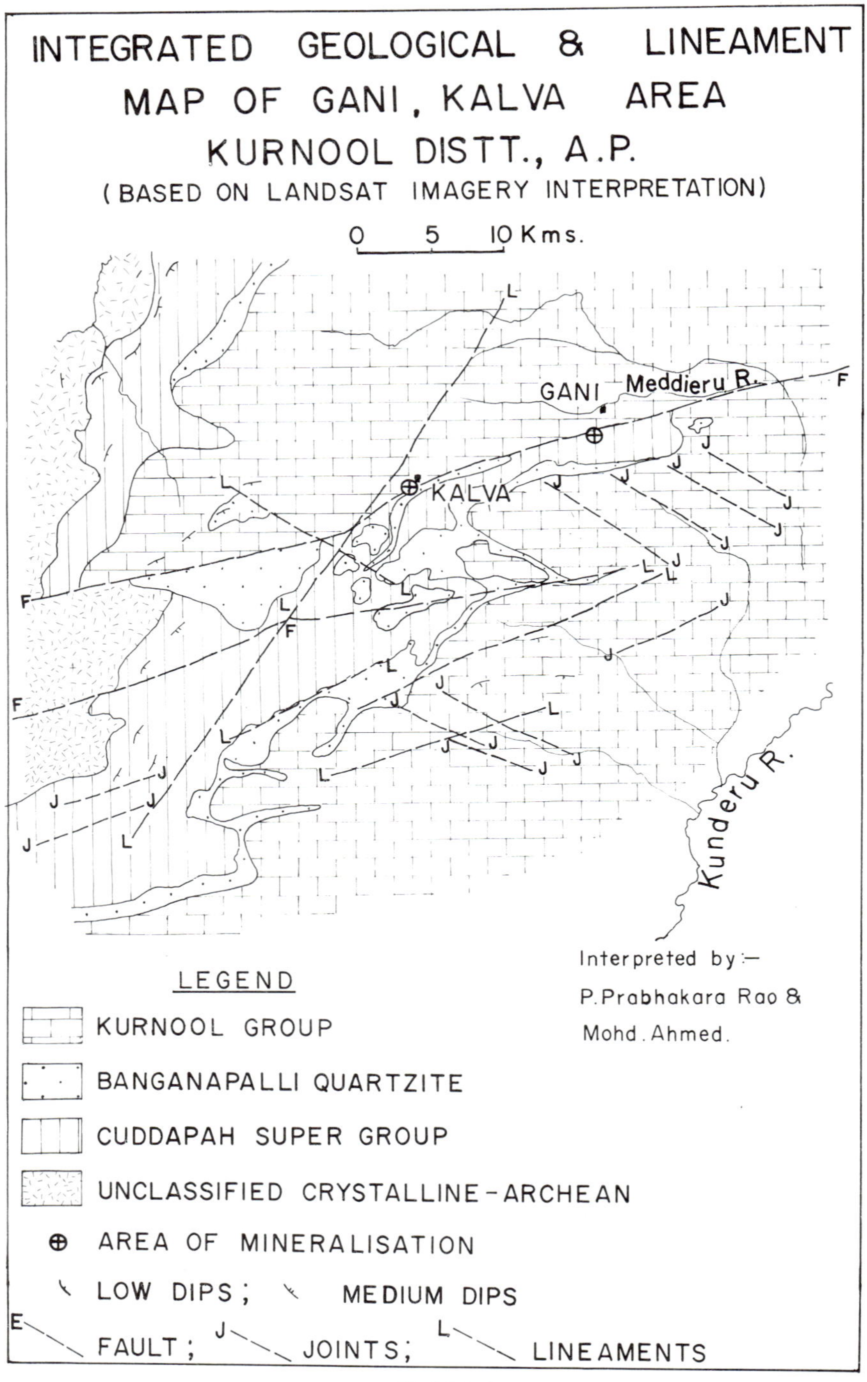
INTEGRATED GEOLOGICAL & LINEAMENT
MAP OF GANI, KALVA AREA
KURNOOL DISTT., A.P.
(BASED ON LANDSAT IMAGERY INTERPRETATION)
0 5 10 Kms.
GANI
Meddieru R.
KALVA
Kunderu R.
F
L
J
Interpreted by:-
P. Prabhakara Rao &
Mohd. Ahmed.
LEGEND
KURNOOL GROUP
BANGANAPALLI QUARTZITE
CUDDAPAH SUPER GROUP
UNCLASSIFIED CRYSTALLINE-ARCHEAN
AREA OF MINERALISATION
LOW DIPS; MEDIUM DIPS
FAULT; JOINTS; LINEAMENTS

FIGURE 5.

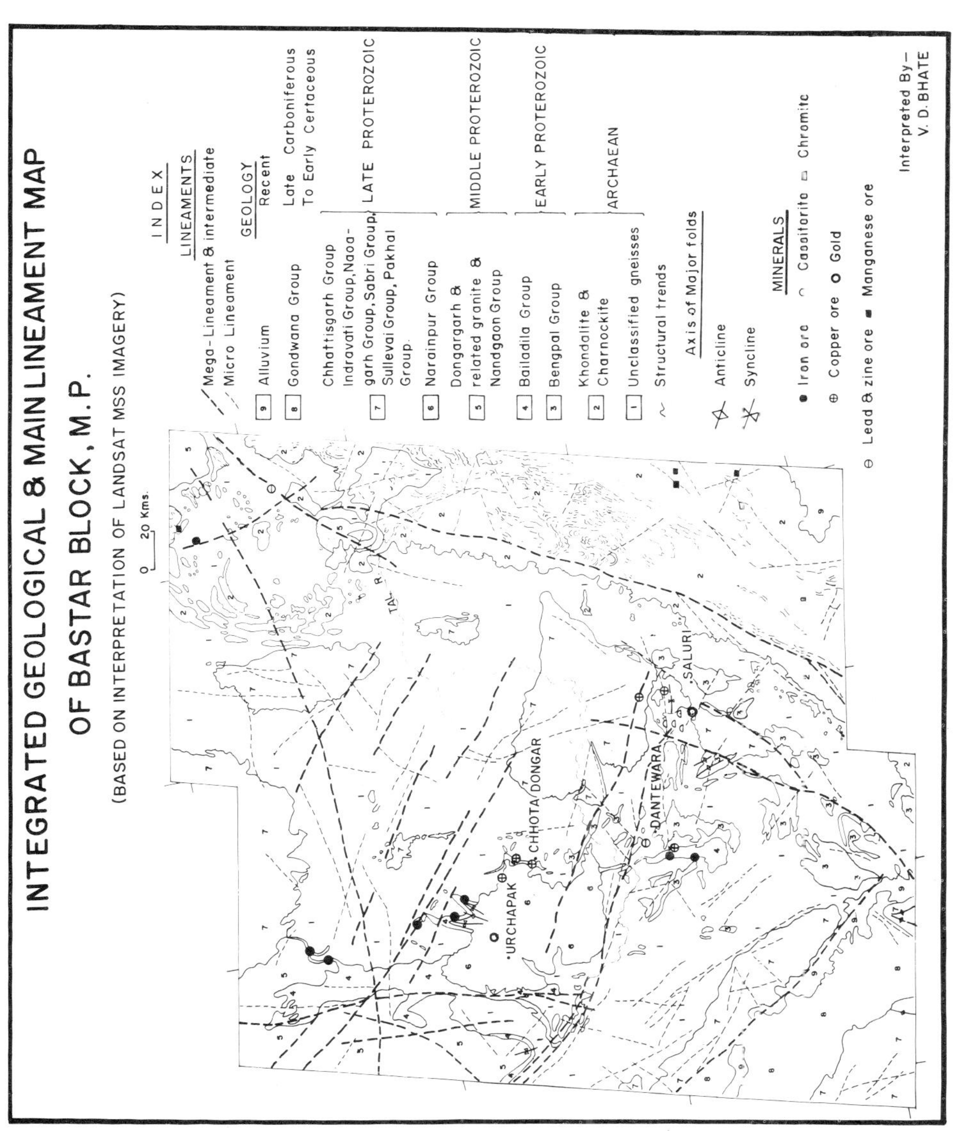
INTEGRATED GEOLOGICAL & MAIN LINEAMENT MAP
OF BASTAR BLOCK, M.P.
(BASED ON INTERPRETATION OF LANDSAT MSS IMAGERY)
0 20 Kms.
INDEX
LINEAMENTS
Mega-Lineament & intermediate
Micro Lineament
GEOLOGY
9 Alluvium
Recent
8 Gondwana Group
Late Carboniferous To Early Certaceous
7 Chhattisgarh Group Indravati Group, Naoagarh Group, Sabri Group, Sullevai Group, Pakhal Group.
LATE PROTEROZOIC
6 Narainpur Group
5 Dongargarh & related granite & Nandgaon Group
MIDDLE PROTEROZOIC
4 Bailadila Group
3 Bengpal Group
EARLY PROTEROZOIC
2 Khondalite & Charnockite
1 Unclassified gneisses
ARCHAEAN
Structural trends
Axis of Major folds
Anticline
Syncline
MINERALS
Iron ore
Cassitorite
Chromite
Copper ore
Gold
Lead & zine ore
Manganese ore
Interpreted By—
V. D. BHATE
URCHAPAK
CHHOTA DONGAR
DANTEWARA
SALURI

FIGURE 6.

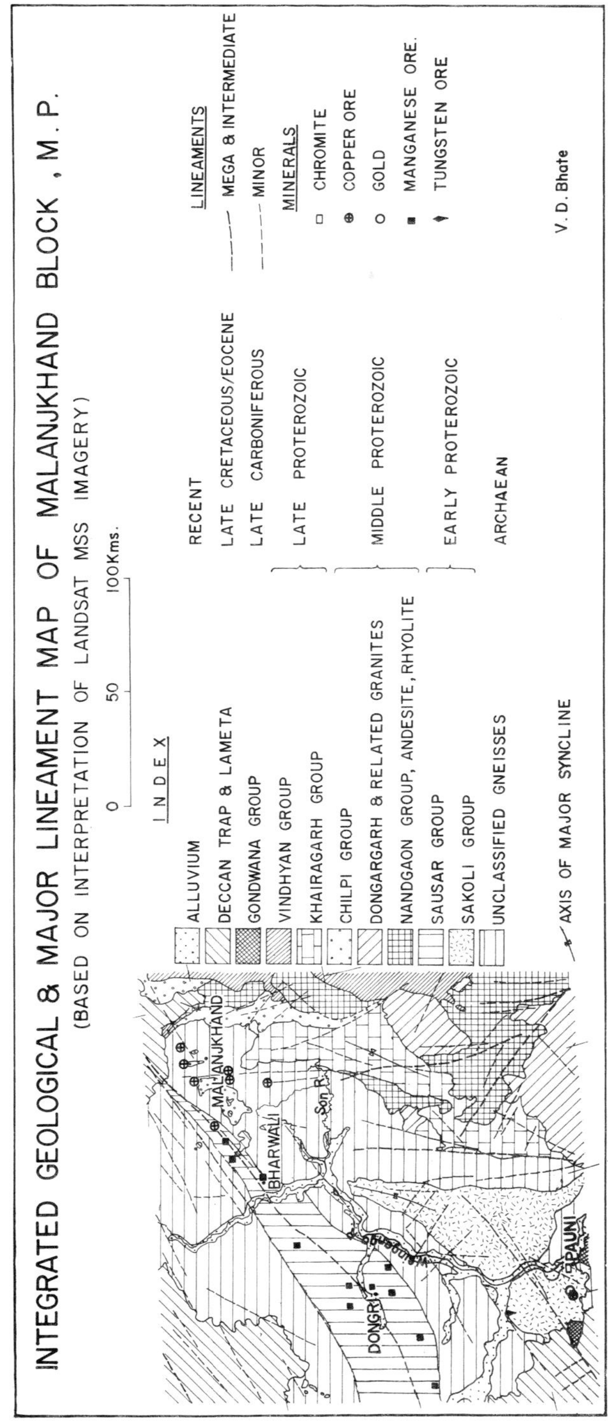
INTEGRATED GEOLOGICAL & MAJOR LINEAMENT MAP OF MALANJKHAND BLOCK , M. P.
(BASED ON INTERPRETATION OF LANDSAT MSS IMAGERY)
0
50
100 Kms.
INDEX
ALLUVIUM
DECCAN TRAP & LAMETA
GONDWANA GROUP
VINDHYAN GROUP
KHAIRAGARH GROUP
CHILPI GROUP
DONGARGARH & RELATED GRANITES
NANDGAON GROUP, ANDESITE, RHYOLITE
SAUSAR GROUP
SAKOLI GROUP
UNCLASSIFIED GNEISSES
AXIS OF MAJOR SYNCLINE
RECENT
LATE CRETACEOUS/EOCENE
LATE CARBONIFEROUS
LATE PROTEROZOIC
MIDDLE PROTEROZOIC
EARLY PROTEROZOIC
ARCHAEAN
LINEAMENTS
MEGA & INTERMEDIATE
MINOR
MINERALS
CHROMITE
COPPER ORE
GOLD
MANGANESE ORE.
TUNGSTEN ORE
V. D. Bhate
MALANJKHAND
BHARWALI
Son R.
DONGRI
PAUNI

FIGURE 7.

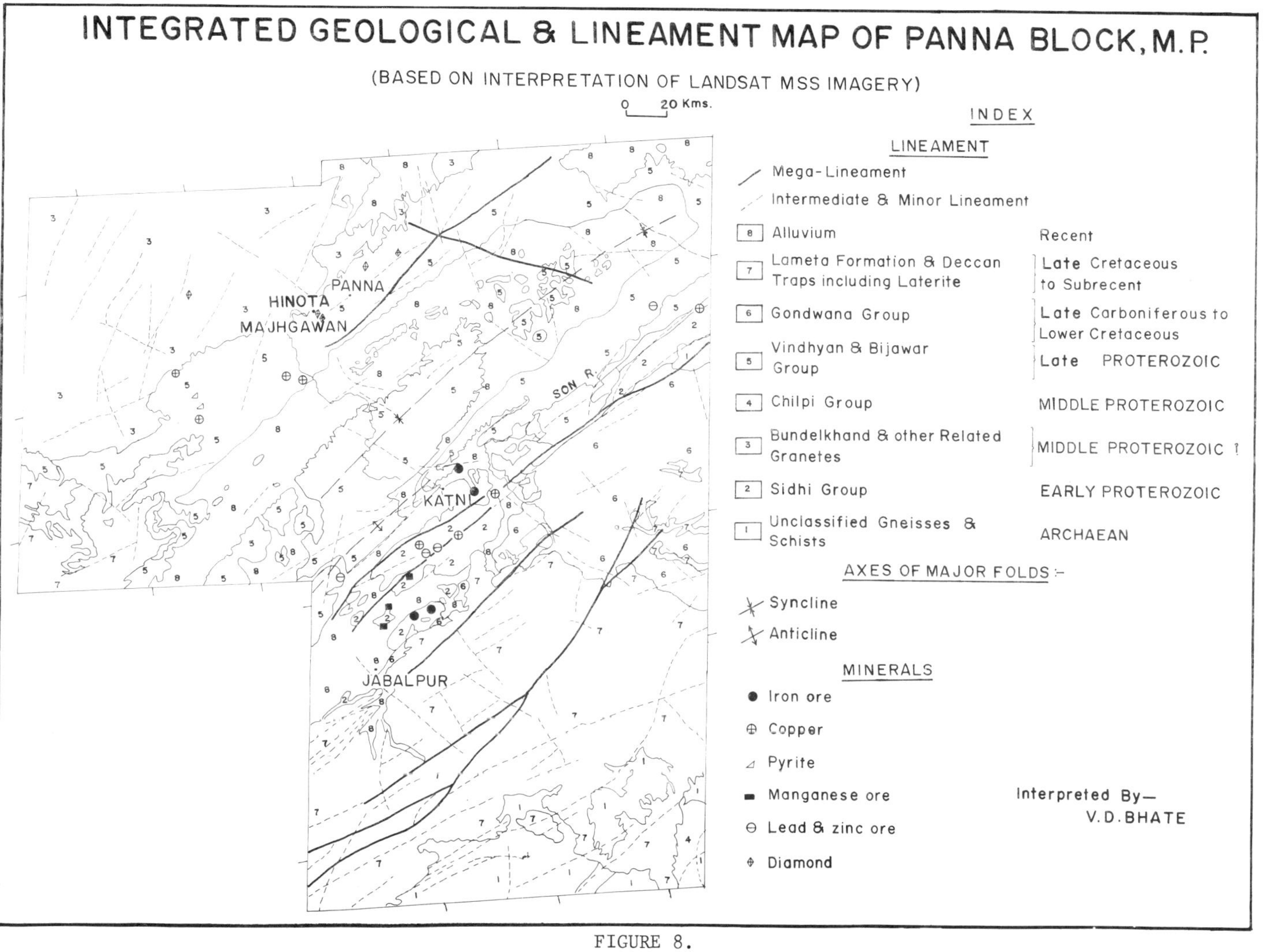
INTEGRATED GEOLOGICAL & LINEAMENT MAP OF PANNA BLOCK, M.P.
(BASED ON INTERPRETATION OF LANDSAT MSS IMAGERY)
0 20 Kms.
INDEX
LINEAMENT
Mega-Lineament
Intermediate & Minor Lineament
8 Alluvium
Recent
7 Lameta Formation & Deccan Traps including Laterite
Late Cretaceous to Subrecent
6 Gondwana Group
Late Carboniferous to Lower Cretaceous
5 Vindhyan & Bijawar Group
Late PROTEROZOIC
4 Chilpi Group
MIDDLE PROTEROZOIC
3 Bundelkhand & other Related Granetes
MIDDLE PROTEROZOIC ?
2 Sidhi Group
EARLY PROTEROZOIC
1 Unclassified Gneisses & Schists
ARCHAEAN
AXES OF MAJOR FOLDS:-
Syncline
Anticline
MINERALS
Iron ore
Copper
Pyrite
Manganese ore
Lead & zinc ore
Diamond
Interpreted By—
V.D.BHATE
PANNA
HINOTA
MAJHGAWAN
SON R.
KATNI
JABALPUR

FIGURE 8.

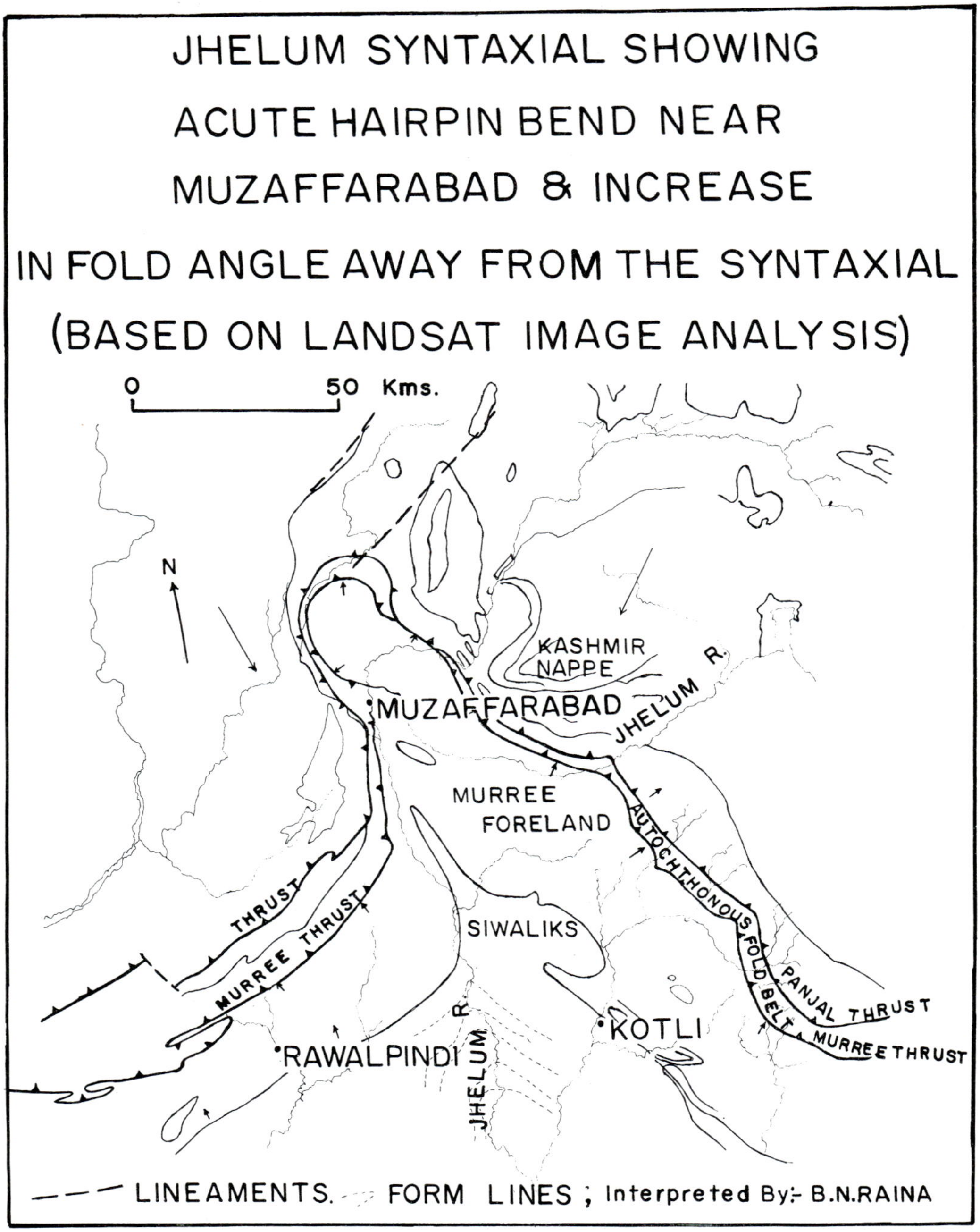

JHELUM SYNTAXIAL SHOWING
ACUTE HAIRPIN BEND NEAR
MUZAFFARABAD & INCREASE
IN FOLD ANGLE AWAY FROM THE SYNTAXIAL
(BASED ON LANDSAT IMAGE ANALYSIS)
0 50 Kms.
N
KASHMIR NAPPE
MUZAFFARABAD
JHELUM R.
MURREE FORELAND
AUTOCHTHONOUS FOLD BELT
THRUST
MURREE THRUST
SIWALIKS
PANJAL THRUST
MURREE THRUST
KOTLI
RAWALPINDI
JHELUM R.
LINEAMENTS. FORM LINES ; Interpreted By:- B.N.RAINA

FIGURE 9.

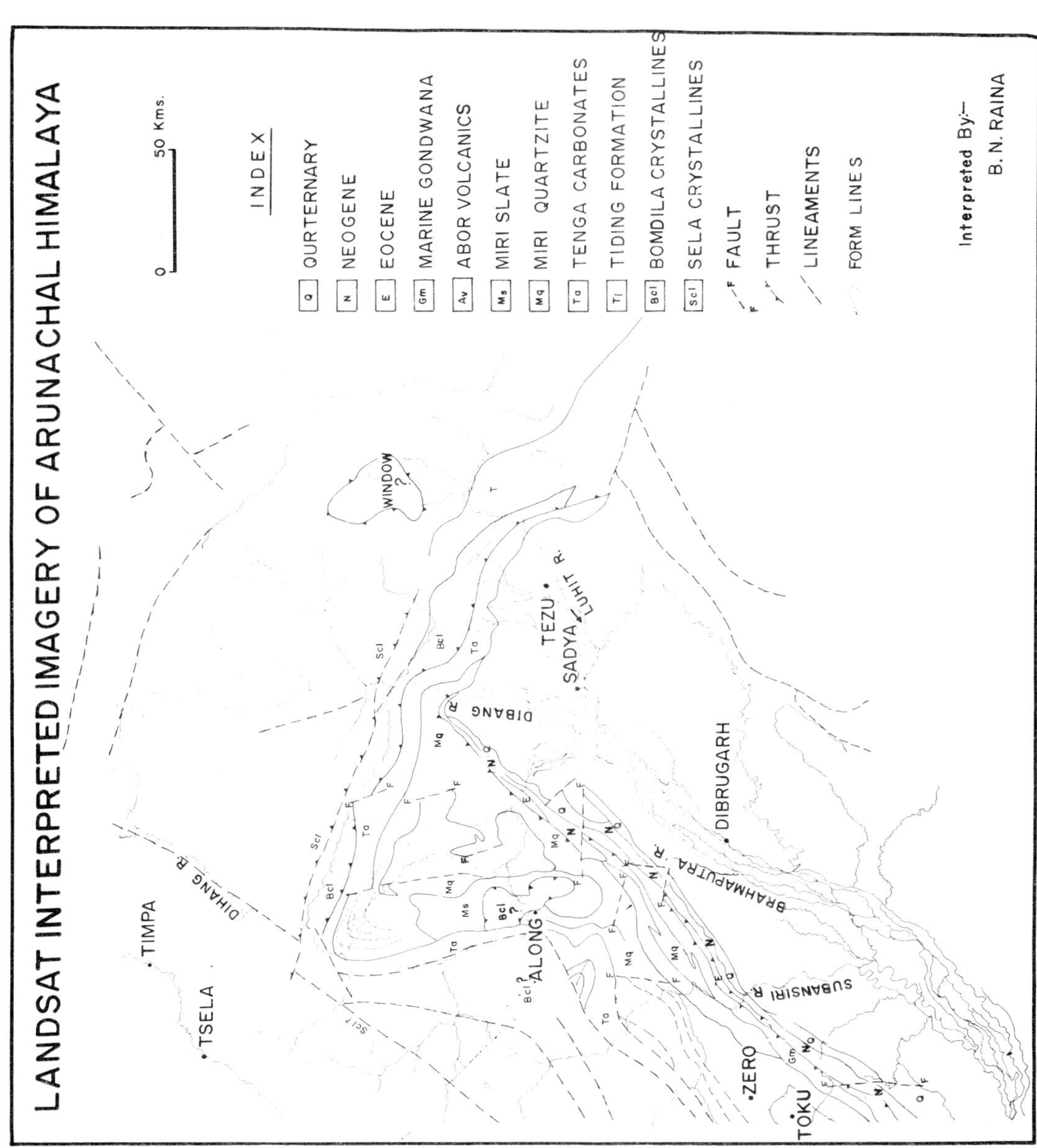
LANDSAT INTERPRETED IMAGERY OF ARUNACHAL HIMALAYA
0 50 Kms.
INDEX
Q QURTERNARY
N NEOGENE
E EOCENE
Gm MARINE GONDWANA
Av ABOR VOLCANICS
Ms MIRI SLATE
Mq MIRI QUARTZITE
Ta TENGA CARBONATES
Ti TIDING FORMATION
Bcl BOMDILA CRYSTALLINES
Scl SELA CRYSTALLINES
FAULT
THRUST
LINEAMENTS
FORM LINES
Interpreted By:-
B. N. RAINA
TIMPA
TSELA
DIHANG R.
ALONG
ZERO
TOKU
SUBANSIRI R.
BRAHMAPUTRA R.
DIBRUGARH
DIBANG R.
SADYA
TEZU
LUHIT R.
WINDOW ?

FIGURE 10.

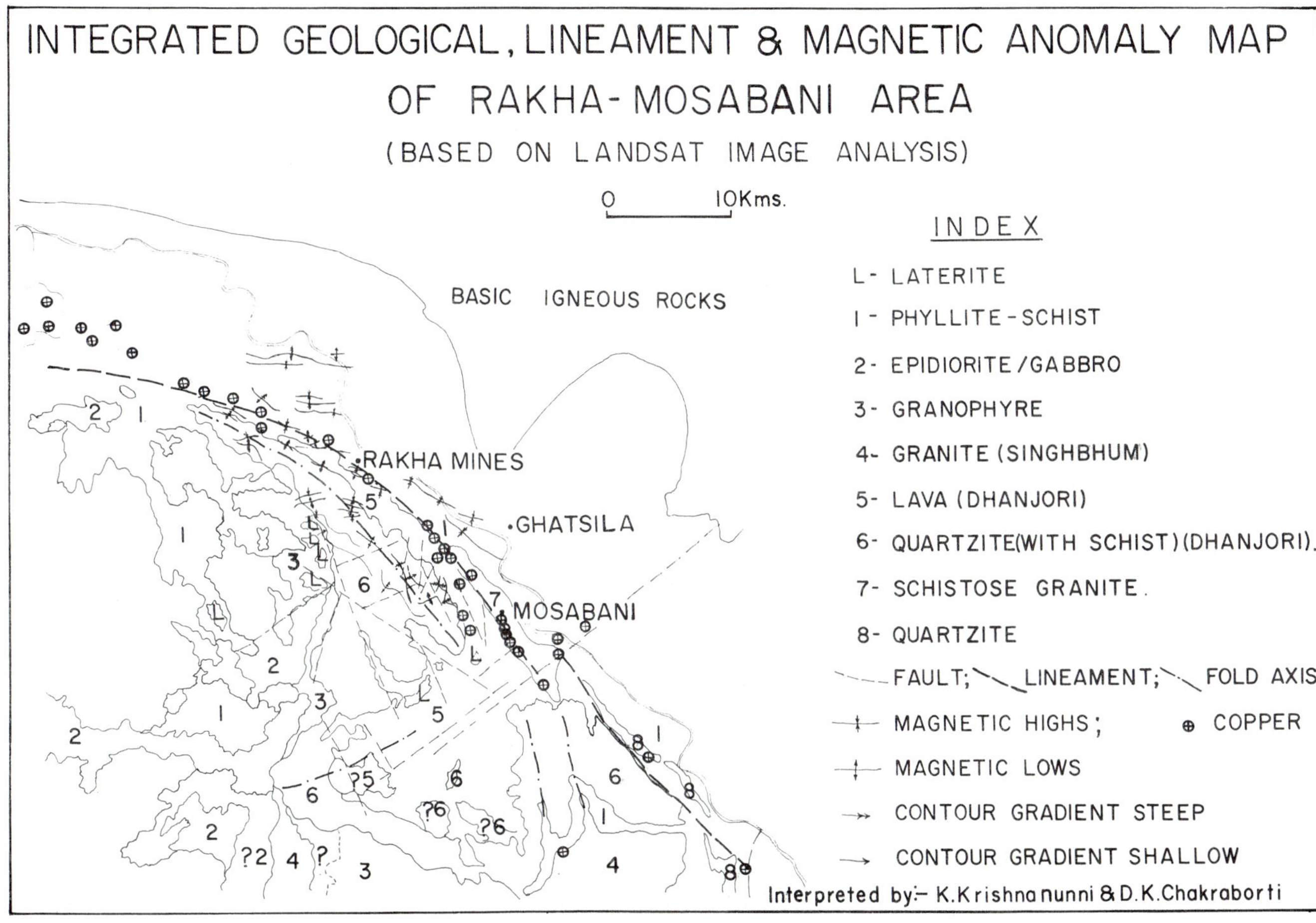
INTEGRATED GEOLOGICAL, LINEAMENT & MAGNETIC ANOMALY MAP
OF RAKHA-MOSABANI AREA
(BASED ON LANDSAT IMAGE ANALYSIS)
0 10Kms.
BASIC IGNEOUS ROCKS
RAKHA MINES
GHATSILA
MOSABANI
INDEX
L - LATERITE
1 - PHYLLITE-SCHIST
2 - EPIDIORITE/GABBRO
3 - GRANOPHYRE
4 - GRANITE (SINGHBHUM)
5 - LAVA (DHANJORI)
6 - QUARTZITE(WITH SCHIST)(DHANJORI).
7 - SCHISTOSE GRANITE.
8 - QUARTZITE
FAULT; LINEAMENT; FOLD AXIS
MAGNETIC HIGHS; ⊕ COPPER
MAGNETIC LOWS
CONTOUR GRADIENT STEEP
CONTOUR GRADIENT SHALLOW
Interpreted by:- K.Krishnanunni & D.K.Chakraborti

FIGURE 11.

FRACTURE MAPPING OF THE NARMADA-TAPTI BASIN USING LANDSAT IMAGERY

S. N. Pandey

Department of Applied Geology, University of Saugar, Sagar (M.P.), India

ABSTRACT

The straight course and the westward flow of the Narmada and Tapti rivers are considered to be the rivers of tectonic significance in the geological history of Peninsular India. In the recent past the Narmada valley region has been considered as 'a line of special significance','a tectonic depression' and a 'rift valley'. However, none of these features have been proved conclusively.

In the present study an attempt has been made to map the fractures and faults of the Narmada-Tapti valley region using Landsat imagery. This brings out very clearly the existing fractures and faults of this region. The result of this investigation indicates that the region between the Narmada and Tapti rivers is highly fractured and faulted. The majority of these linear elements are oriented in the ENE-WSW direction, parallel to the course of the Narmada river, indicating the tectonic influence of the rocks over the geomorphology of this region.

INTRODUCTION

There are two rivers in the west central India, namely Narmada and Tapti, which have attended great geological significance from the morpho-tectonic point of view. Certain unusual features of these rivers such as their long straight courses, westward flow and sharp contrast in the geology, structures and geomorphology of the Narmada river have attracted the attention of geologists for the the past few decades.

The area under consideration for fracture mapping falls mainly between the Narmada and Tapti rivers involving a major part of the Satpura Plateau. The course of the Narmada river in the present investigation extends from the proposed Navagam dam-site in the Gujarat state to the Jabalpur town in the Madhya Pradesh state.

The main geological formations in geochronological order are the Archaeans (Crystallines and Gneisses), Dharwars (Pre Cambrian meta-sediments), Vindhyans (unmetamorphosed sediments), and Deccan traps and alluvium (Fig.1).

The physiography of the area is most clearly depicted on the physiographic map published by the National Atlas of India (Scale 1:1,000,000). The present area incorporates two of the ancient mountains, namely the Vindhyas and Satpuras which are the outstanding features of the landscape.

PURPOSE AND SIGNIFICANCE OF PRESENT INVESTIGATION

The elongated zone of the Narmada valley was considered to be a 'line of special significance',(1), a 'rift valley' and a'faulted depression'(2). These views, however, have been inconclusive due to inadequate supporting evidence. The geological work done by the earlier investigators had their own limitations of available base material.

A systematic regional coverage carried out with the Landsat imagery enabled the author to undertake this task on an individual basis. In the present investigation, the Landsat imagery depicts the regional picture of the Narmada-Tapti valleys revealing the existance of fractures and faults, the true orientation of the main rivers and their tributaries, and other morphological evidences to assess the tectonic nature of the terrain.

TECHNIQUE OF INVESTIGATION

The present fracture mapping has been carried out following Kaiser (3) and Blanchet (4). The term macro-fracture has been used, as on such a small scale satellite imagery, only objects of large dimensions ranging from a few hundred meters or more can be identified. The other linear structural elements such as micro-fractures, fissures etc. do not fall within the resolution range of the image.

On the Landsat imagery of the Narmada-Tapti valley region, the macro-fractures are reflected in many forms like straight fracture valleys, linear topographic features, straight segments of river and stream courses, and abrupt linear tonal variations. The general objective of the fracture mapping is to relate such shallow to deep seated structures and geologic anomalies to the major tectonic episodes of the area.

INTERPRETATION OF LANDSAT IMAGERY

In the present investigation the MSS Band 6 was preferred because it provides the best information on the geology, geomorphology, geologic structures, soil, soil-moisture and land-water contrast of the region.

The 9"X9" photo-prints of the Landsat imagery were studied under the mirror stereoscope which permitted a better perspective of the terrain under a high magnification. In this way, all the traceable macro-fractures and faults were marked on the Landsat prints and later transferred on a base-map on the scale of 1"=16 miles (1:1,000,000). The true courses of the Narmada and Tapti rivers and their tributaries were also traced along with the significant geomorphic features (Fig.1).

The morpho-tectonic characteristics of different regions as revealed from the study of the Landsat imagery are as follows:

Satpura Plateau

To the south of the Narmada river, the Satpura Range, essentially a lava capped plateau of Archaean gneisses and schists, abuts against steep sided hills of Gondwana sandstones. The trends of the ancient folds are in the ENE-WSW direction to which conform the present topographic features. The hills of the western Satpura have been considerably affected by orogenic disturbances as indicated by the tilting of the trappean beds and presence of regional faults and fractures. The eastern Satpura is comparatively free from structural disturbances and is partly developed on the Upper Gondwana sandstone and partly on the Trap rocks.

The upland area between the Narmada and Tapti valley is characterised by fracturing and regional faulting. On the Landsat imagery the faults are recognised mainly by the following four criteria:

(a) Sharp, straight and angular scarpments facing the river valley.
(b) Truncation of macro-fractures along a straight or a slightly curved line.
(c) Truncation of fractured blocks against unfractured blocks, and
(d) Lateral displacement of rock fractures or escarpments.

Some of the important observations made from the Landsat imagery of the Satpura Plateau are as follows:

(i) The results of the fracture mapping of this region makes it clear that the western Satpura is highly fractured and faulted with their strike orientation mainly in the ENE-WSW direction. Another set of fractures and faults are oriented in the NW-SE and NNW-SSE directions (Fig.2-A, point 'FR').

(ii) The density and distribution of fractures and faults is much more on the western Satpura as compared to its eastern parts. On this part of the plateau the fractures have cut across all the geological formations, i.e. the Crystalline Gneisses, the Gondwanas and the Deccan traps. In the whole area, the maximum density of fractures has been determined around the proposed Navagam dam-site, Pandey (5). On the eastern Satpura (south of Narsinghpur and Jabalpur), the effect of fractures on the topography has become so weak that they have hardly influnced the courses of rivers. The course of the Narmada river in this part of the country has practically no influence of the ENE-WSW oriented macro-fractures (Fig.2, point 'N').

(iii) Another higher density of fractures is confined in the region to the south of Harda and Hoshangabad. This region clusters the ENE-WSW oriented fractures and dykes intersected by the NNW-SSE oriented fracture sets (Fig. 2-C, points 'FR').

(iv) One of the very remarkable facts noticed on the satellite imagery is the unusual straightness of the northern confines of the Satpura Plateau with that of the Narmada alluvium and other older rocks. This straight contact clearly continues along Khandwa, south of Hoshangabad, Piparia, Narsinghpur and Jabalpur. Further west of Khandwa, this contact is expressed by a sharp tonal contrast along a line. This contrast gradually dies down to the south and south-east of Khargon (Fig.2, points 'C','D',&'E'). All these criteria are indicative of a regional fault which runs for about 260 miles (418 km) in the present area.

Narmada Valley

The Narmada river flows through an asymmetrical valley-section between the Vindhya Range on the north and the Satpura Range on the south. From Jabalpur westward, the river first flows through the gentle, undulating open plain of Hoshangabad. To the north of Harda, the river leaves the alluvial plain and flows through the quartzite hills and the granitic gneisses. Further west, the river flows through the gorges which have been carved out of Vindhyan sandstones. From Barwah westward the river flows almost in a straight course with a number of steep angular turns up to Navagam. The main characteristic feature of the Narmada river is its flow through straight segments with sharp angular turns. The following observations have been made from the study of the Landsat imagery:

(i) The course of the Narmada river and its tributaries between Hoshangabad and Jabalpur are characterised by smaller straight river segments with sharp angular turns and junctions. This may indicate the initial control of the intersecting fractures on the courses and orientations of the rivers, before this tract of the valley got filled up with alluvium (Fig.2-E, point 'A').

(ii) From Barwah to Navagam, the river follows a straight course forming deep gorges for more than 180 miles (290 km). The river as a rule always turns at sharp angles. Throughout the course of the river, there is no indication of meandering or river deposits. These evidences strongly indicate the fault and fracture control on the course of the river (Fig.2, point 'S').

(iii) A remarkable point to note is that at a point east of Barwani, the ENE -WSW course of Narmada is transacted by a NNW-SSE running fault which has brought the corresponding sharp angular turns in the course of the river (Fig.2-A, point 'T'). Further west of this point, as is evident on the Landsat imagery, the river valley has an increasing density of fractures up to the Navagam dam-site area.

(iv) The region to the north of Narmada has clear indications of faulting in the NNW-SSE direction showing lateral displacement. This faulting has also involved crystalline metamorphics of Chhota Udaipur area (Fig.2-A, point 'L').

Tapti Valley

The Tapti river flows to the south of the Satpura Range towards the west through the wide plains of Khandesh in the Maharashtra Plateau Province. Unlike the river Narmada, the river Tapti hardly shows any control of faults and fractures. The river has a typical dendritic pattern with a few tributaries. The following observations have been made from the Landsat imagery:

(i) The scarpments bounding the northern portion of the river are straight and angular. They are indicative of faulting. The straight scarpments invariably meet at sharp angles. However, this scarp-line is also the line of truncation of fractures dominant on the northern Satpura upland (Fig.2-A).

(ii) The entire course of the Tapti river and its tributaries flow through the topographic depression, now defining the Tapti valley.

(iii) The orientation and nature of the Tapti river segments do not show any indication of faulting and fracturing. Even if such regional structures had controlled the drainage in the initial stage of its development, they might have vanished due to deep erosion along the valley and their traces would have been covered under the thick soil cover and alluvium (Fig.2-A).

CONCLUSION

The visual interpretation of the Landsat imagery of the Narmada-Tapti valley region has led to the following broad conclusions:

1. In the geology of Central India, the Narmada-Tapti zone shows abundant fracturing and faulting. The fractures and faults are oriented mainly in the ENE-WSW direction parallel to the course of the Narmada and Tapti rivers. This set of fractures is intersected by another fracture set oriented mainly in the NNW-SSE direction.

2. The maximum density of fractures is concentrated in the westernmost part of the Satpura Range closer to the Narmada river where the proposed sites of the high Navagam dam are located.

3. The Satpura Range on its western and southern parts displays a number of faults oriented in the ENE-WSW direction. Some of the faults of relatively smaller magnitude are oriented in the NW-SE and NNW-SSE direction. Their occurrence is confined to the west of Barwani and Khargon and south of Khandwa.

4. The course of the Narmada river between Navagam and Barwah is dominantly controlled by the faults and fractures along the river valley. A certain amount of fracture control is also noticed in the eastern portion between Hoshangabad and Jabalpur, although this tract of the valley is occupied by alluvium.

5. The Landsat imagery has proved extremely useful in delineating the faults and fractures of the Narmada-Tapti valley area. It is hoped that the morpho-tectonic map so prepared will be an useful base map for those who wish to explore some part of this region in greater details using aerial photographs and fieldwork.

ACKNOWLEDGEMENTS

The author is grateful to the Association of Commonwealth Universities, London (U.K.) for providing funds and necessary help to procure the Landsat imagery of Central India from U.S.A.

REFERENCES

1. W.D. West, Current Science 143-144, 31 (1962).
2. J.B. Auden, Proc. Nat. Inst. Sci. Ind. 315-340, 15 (1949).
3. E.P. Kaiser, (Abs) Bull. Geol. Soc. Amer. 1475, 12 (1950).
4. P.H. Blanchet, Bull. Amer. Assoc. Petrol. Geol. 1748-1759, 41 (1957).

5. S.N. Pandey, Proc. Int. Symp. Res. Engg. Tech. II, I.I.T. Bombay (Jan. 8-11, 1979).

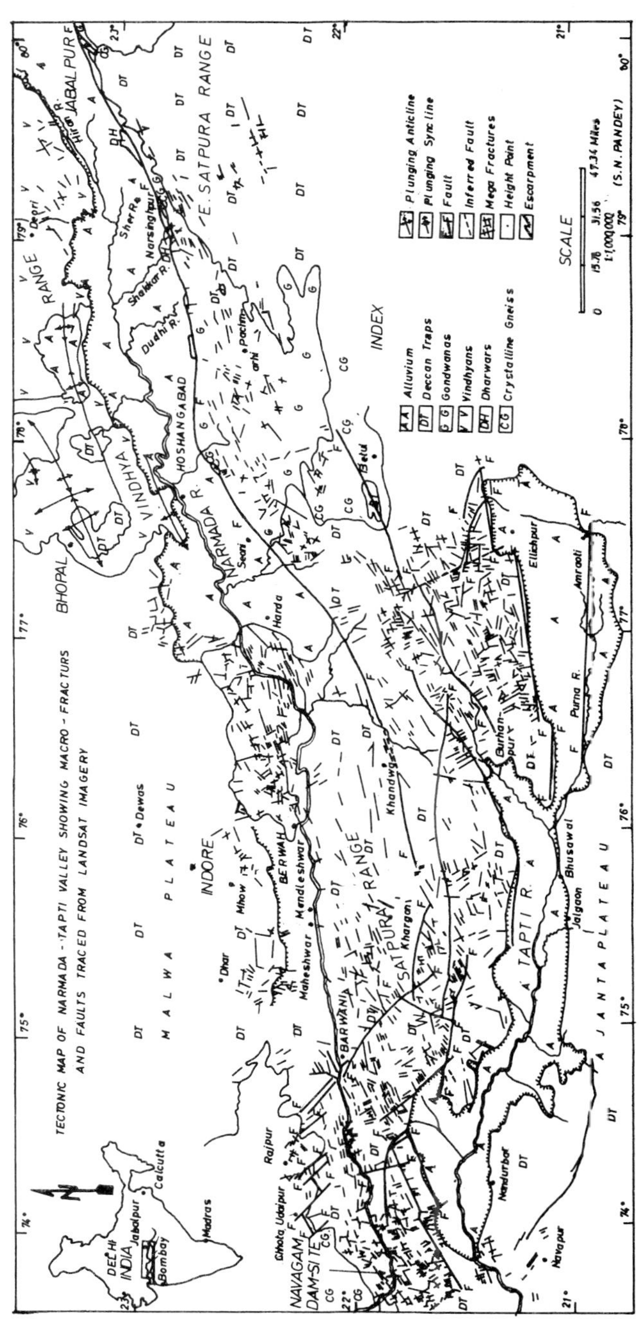

Fig.1 Tectonic Map of Narmada-Tapti Valley Showing Macro-fractures and Faults Traced from Landsat Imagery.

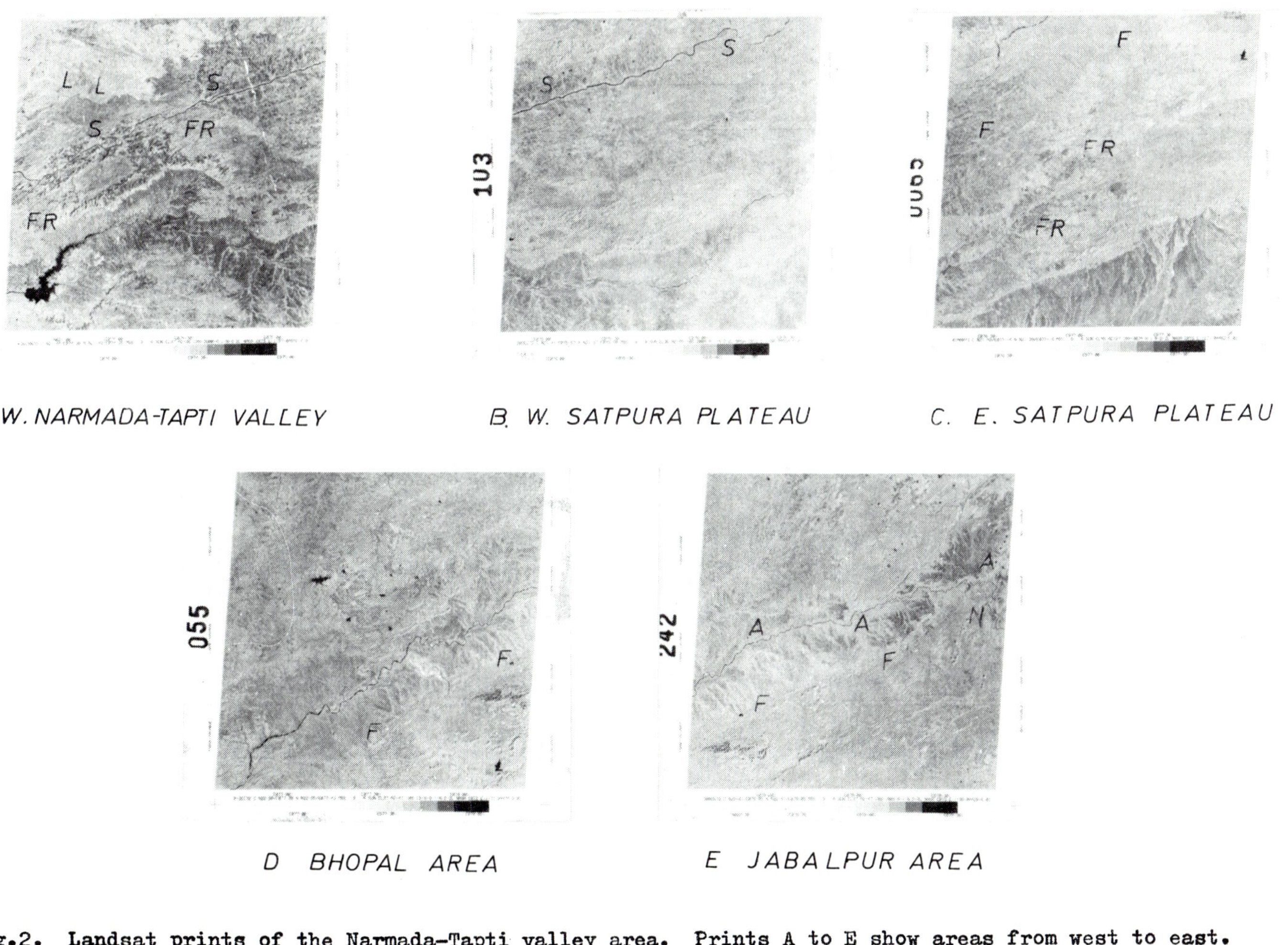

Fig.2. Landsat prints of the Narmada-Tapti valley area. Prints A to E show areas from west to east. (Prints prepared from Landsat negative transparency).

LINEAMENT STUDY OF THE BASTAR DISTRICT, MADHYA PRADESH, INDIA, FROM LANDSAT IMAGERY

V. D. Bhate

Geological Survey of India

ABSTRACT

Study of the Landsat-1 MSS imagery covering the Bastar district, Madhya Pradesh, India, has brought out the lineament pattern, major structural trends, and main geological formations of the area. The lineaments in Bastar trend mainly NW-SE, NE-SW, N-S, NNE-SSW, E-W and ENE-WSW, the NW-SE trend being the most predominant. The study of the lineaments indicates that the Central Bastar Plateau probably represents an uplifted block bounded on three sides by major faults.

INTRODUCTION

The Bastar district of Madhya Pradesh is one of the test sites selected for study under the IGCP Project No. 143 - Remote Sensing and Mineral Exploration. In this connection four Landsat frames were analysed covering the entire Bastar district, parts of Raipur, Durg, Rajnandgaon districts of Madhya Pradesh and adjoining parts of Maharashtra, Orissa and Andhra Pradesh, India. The study area lies between the geographic coordinates N Lat. 18°00' to 20°45' and E Long. 80°00' to 83°30'.

The visual interpretation involving four Landsat frames was carried out by monoscopic study of the individual band information as well as combinations of two bands under the stereoscope. The results of this preliminary study are presented in this paper.

REGIONAL GEOLOGICAL SETTING

The study area forms a part of the northern extension of the Indian Peninsular shield and is predominantly occupied by Precambrian rocks, except for a narrow NW-SE trending belt of the Paleo-Mesozoic Gondwana sediments along the Godavari valley.

The Precambrian of this region includes a gneissic complex; Charnockite and Khondalite suite of rocks; metasediments of the Bengpal and Bailadila Groups; acid and basic volcanics of the Nandgaon Group; Dongargarh granite; gabbro-anorthosite complex; volcano-sedimentary sequence of the Narainpur Group; and the sediments of the Pakhal, Chhattisgarh, Indravati and Sullavai Groups [1, 2, 3, 4, 5, 6 & 7]. The chrono-stratigraphic relationship among the various groups of the early Precambrian is still to be worked out satisfactorily and remains a matter of controversy due to the inherent difficulties in classifying these ancient rocks which have a complex evolutionary history.

INTERPRETATION RESULTS

Geology

Identification of only broad lithological units has been possible from the Landsat imagery (Fig. 1). The band 5 and 7 images were found most suitable for delineation of the lithological boundaries and structural features.

The area occupied by the Charnockite and Khondalite Group of rocks is characterised by a bold relief and highly dissected topography. The discrimination of Charnockites and Khondalites is rather difficult on the imagery due to their similar photo characters. The folded rocks of Bengpal and Bailadila Groups with linear quartzite ridges constituting a prominent topographic feature, are very well reflected on the imagery. However, in areas of low relief, separation of these rocks from the surrounding granitic gneisses is not possible due to lack of clear tonal differences and topographic expression. The meta-andesites and meta-rhyolites of Nandgaon Group can be easily discriminated from the Dongargarh granite on the basis of distinct tonal differences.

A large oval patch, exhibiting anomalous dark tone in all the four band images, has been observed in the Tel river basin (Fig. 1). It has been provisionally interpreted as a gabbroic body and is grouped with the Gabbro-Anorthosite complex on the basis of the occurrence of a minor anorthosite body shown in the published geological map.

The volcano-sedimentary sequence of Narainpur Group, which has not been documented so far in available published geological maps, has been very well brought out on the imagery. It has been delineated with fair degree of accuracy on the basis of distinct topographic expression and alternating dark and light tones exhibited by the basic flows and sandstone beds respectively. A small patch of similar sequence of rocks with identical photo characters has been recognised in Tulsi Dongar area to the south of Indravati basin. The study of aerial photos of Tulsi Dongar area has also confirmed this interpretation which has to be verified by field examination.

The flat lying sediments of Chhattisgarh, Indravati and Sullavai Groups and the folded sediments of Pakhal Group could be easily identified on band 5 images. The Upper and Lower Gondwana of Godavari valley could not be delineated with confidence. Similarly, the southern contact of Gondwana with Pakhal is not clearly discernible on the imagery. The Quaternary deposits exhibiting very light tone were delineated along the Godavari valley.

The prominent NE-SW Eastern Ghat trend in the Charnockite and Khondalite terrain is very well reflected on the imagery. The gradual swerve in the strike of the structural trend in these rocks in the eastern part of the area is also clearly displayed. Some of the major folds with NE-SW, NNE-SSW and N-S axial trends can also be recognised in these rocks. In the Bengpal and Bailadila metasediments mega-folds with NW-SE, N-S and NE-SW axial trends can be made out on the imagery. In the megascopic scale, the Bailadilas appear to have been folded into open antiforms and tightly appressed synforms. The Pakhal sediments exhibit folds with NNE-SSW and NW-SE axial trends.

Lineaments

The lineaments seen in the imagery have been classified into three categories on the basis of their linear extent. The lineament extending for a length of more than 300 km are classed as mega-lineaments those extending from 100 to 300 km as intermediate lineaments and those measuring less than 100 km in length as micro-lineaments. However, for convenience of depiction on the accompanying lineament map (Fig.2), the mega- and intermediate lineaments have been grouped together. The mega- and intermediate lineaments are considered to represent major faults in the area. The micro-lineaments on the other hand, mostly reflect the master joint and fracture pattern of the area. Some of the lineaments corrospond with the known faults in the area.

The major lineament trends are NW-SE, NE-SW, NNE-SSW, E-W, N-S and ENE-WSW. The southeastern part of the area shows a predominance of NE-SW trending lineaments which are probably correlatable to the Eastern Ghat trend, while in the rest of the area the NW-SE trending lineaments are conspicuously developed. These can perhaps be correlated to the NW-SE Godavari-Mahanadi trend. The basic dykes and quartz reefs traversing the granite gneisses and metasediments are also dominantly oriented along the NW-SE direction. The significant lineaments are described briefly below :

The E-W to ENE-WSW trending mega-lineament across the northern part of the area is traceable intermittently from Muramgaon to Bolangir. The straight running course of the upper reaches of Mahanadi and one of its east flowing tributaries are controlled by this lineament. The cliff section of the Keskal ghat located to the south of this lineament perhaps represents a retreated fault scarp related to vertical movement along this lineament.

In the west-central Bastar three subparallel NW-SE trending lineaments constitute a suite of transverse faults that have dislocated the N-S trending structures of Bailadila Group of rocks in Rawghat-Antagarh region. Part of these faults are occupied by silicified zones.

A curvilinear lineament with a general WNW-ESE trend, traceable from Mundaval to Surajgarh in the southern part of Bastar, seems to have affected the folded Bailadila Group of rocks by lateral displacement with downthrow towards south in the Indravati river section in the east and by an easterly drag in the Bhamragarh area in the west. Another parallel lineament, traceable from north of Darba to Dingri, appears to be a reverse fault with its downthrow to the north.

An almost N-S trending lineament partly controlling the straight course of Kotri river has been detected in the western part of the area. This lineament represents the southern extension of the great Darekasa fault of Sarkar [4] in the Dongargarh region.

Two parallel lineaments trending NW-SE have been detected in the Godavari valley. The lineament running along the Godavari river represents the faulted contact between the Pakhal and Gondwana while the other brings the Pakhal in juxtaposition with the basement granite gneisses.

The eastern and the western margins of the Sabri basin are marked by two converging lineaments. A prominent NE-SW trending lineament, possibly related to the Eastern Ghat orogeny, is observed in the southeastern part of the area. The straight running course of Tel river obviously indicates a lineament control.

DISCUSSION

As is well known, mega-lineaments are manifestations of deep-seated fractures within the crustal shell of the earth and as such, are the potential loci of diastrophic movements and epigenetic mineralisation. The Bastar and adjoining areas which form a part of the Indian Peninsular shield, have undergone several episodes of crustal movements and bear ample imprints of these in the form of pronounced orogenic trends e.g. the Eastern Ghat trend, the Godavari-Mahanadi trend, Bailadila trend etc. The surface expression of such orogenic trends are clearly manifested in the form of mega-lineaments, as has indeed been detected in the Landsat imagery. The interpreted lineament map (Fig.2) shows three major trends of the lineaments in the study area (i) the NE-SW, (ii) NW-SE and (iii) N-S. The NE-SW trending lineaments are parallel to the Eastern Ghat orogenic trend and appear to be the oldest of the three, while the NW-SE trending ones, parallel to the Godavari-Mahanadi trend, transecting the Eastern Ghat trend and offsetting the N-S trending Bailadila trend are presumed to be the youngest. The N-S trending lineaments parallelling the axial trend of the mega-folds of the Bailadila metasediments, have a preferential development in the western part of the area studied and appear to be temporally intermediate between the other two lineaments. However, though these mega-lineaments have been tentatively correlated to the major orogenic trends in this part of the Indian shield, they need not be precisely coeval to the respective orogenies, since these lineaments might be manifestations of reactivation of the pre-existing deep-seated weak zones in response to the interplay of later tectonic forces.

The most significant observation regarding mega-tectonics of the area studied is that the Central Bastar Plateau forms an uplifted block (horst) bounded by the (i) curvilinear ENE-WSW trending Muramgaon-Bolangir mega-lineament in the north with downthrow towards north, (ii) N-S trending Kotri mega-lineament in the west with downthrow towards west and (iii) WNW-ESE trending Mundaval-Surajgarh mega-lineament in the south with downthrow towards south. The Muramgaon-Bolangir lineament shows differential vertical displacement in different parts along its length and, as such, has been interpreted as a wrench fault. Uplift of the Central Bastar Plateau involving dislocation of even the younger Indravati-Chhattisgarh sediments, can obviously be dated to post Proterozoic times.

The reported occurrences of mineral deposits e.g. copper, gold and cassiterite in the study area when plotted in the lineament map (Fig.2), lie either in close proximity or on some of the lineaments. The significance of such spatial relationship is quite obvious. However, precise knowledge of the control of localisation of mineral deposits by the lineaments can only be obtained by thorough ground investigation. The recorded gold occurrences in the study area, are of alluvial type, but in view of their situation on mega-lineaments, it is worthwhile to search for their primary source within the lineament zone itself. The NNW-SSE trending cassiterite-bearing pegmatites in the Kudripal-Govindpal-Mundaval area in the southern Bastar, lie within the Mundaval-Surajgarh lineament zone and have been postulated by Bose et.al. [8] to be genetically related to the Darba granite, which also lies in close proximity of this lineament. However, it is yet to be established as to whether this lineament had any control in the emplacement of the Darba granite and its associated cassiterite-bearing pegmatites. In case such a relationship could be established by field examination, the contiguous areas lying along the western extension of this lineament zone and having similar geological set up would be the potential areas for locating more cassiterite-bearing pegmatites.

CONCLUSIONS

The broad synoptic view offered by the Landsat imagery gives a comprehensive regional geotectonic picture. The satellite imagery is also a useful aid in reconnaissance mapping and in updating the existing small scale maps of remote, forest covered, inaccessible regions such as the Bastar area.

The present study though of a preliminary nature has clearly brought out the lineament pattern and other broad structural features of the Bastar area and has indicated the possible relationship of some of the lineaments with known mineral occurrences. Detailed analysis of some selected frames utilising image enhancement techniques will possibly provide further useful data.

ACKNOWLEDGEMENT

The author is thankful to the Director General, Geological Survey of India, for permission to present this paper at the COSPAR 1979 Workshop on Remote Sensing and Mineral Exploration.

REFERENCES

1. H. Crookshank, Mem. Geol. Surv. Ind. 87 (1963)
2. V.D. Bhate and S.V.G.K. Rao, Abstracts, Seminar on Archaeans of Central India, G.S.I., Nagpur (1976)
3. P.K. Ghosh, Rec. Geol.Surv.Ind. 75, Prof. paper 15 (1941)
4. S.N. Sarkar, Jour.Sci.Eng.Res.IIT, Kharagpur, I and II (1957-58)
5. S.M. Dutta, et.al., Abstracts, Seminar on Archaeans of Central India, G.S.I., Nagpur (1976)
6. N.V.B.S. Dutt, Jour. Geol. Soc. Ind. 4 (1963)
7. N.V.B.S. Dutt, Rec. Geol. Surv. Ind. 93 pt.2 (1964)
8. S.K. Bose, et.al., Proceedings, International Symposium on Resources Engineering & Technology, IIT, Bombay, II (1979)

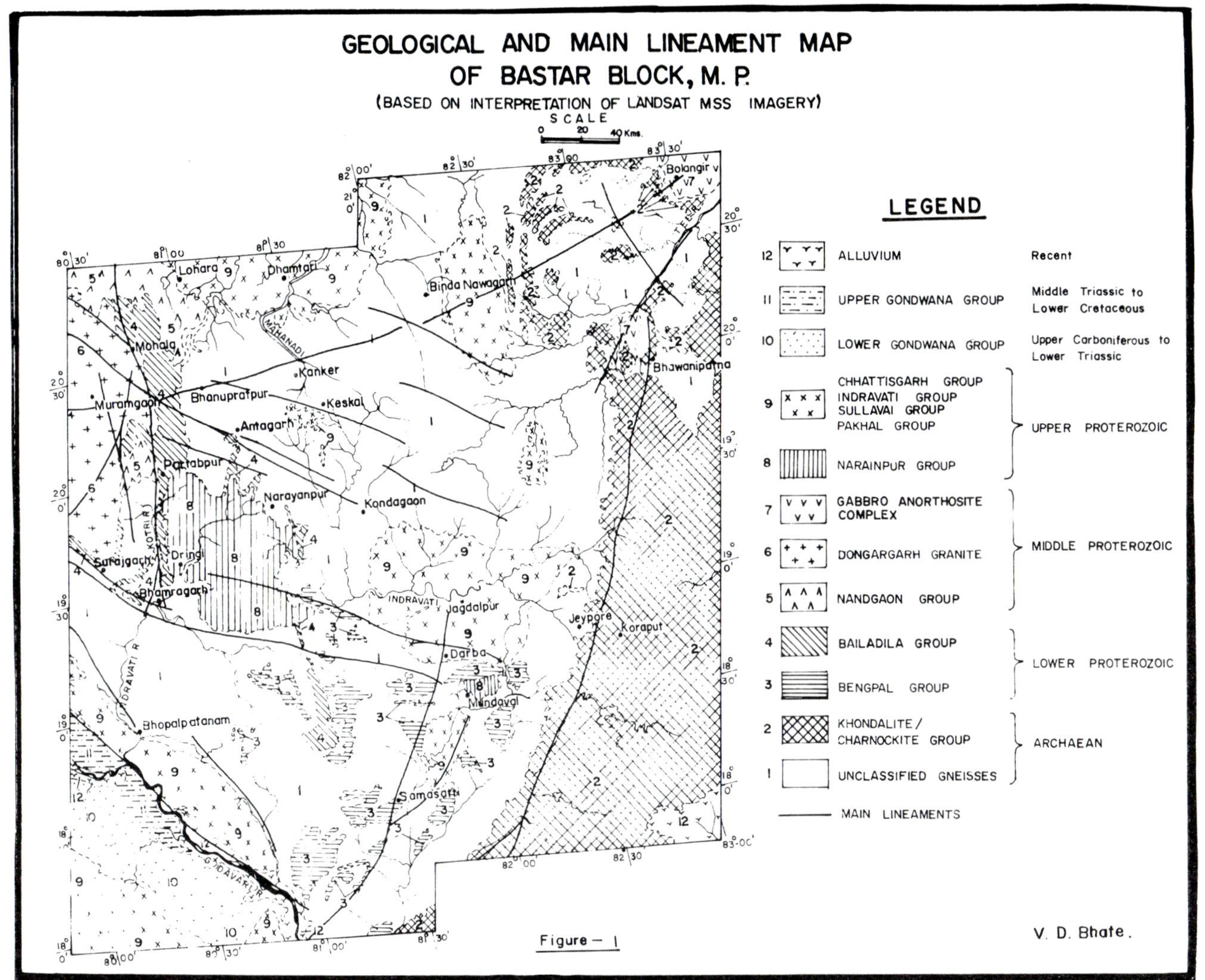
GEOLOGICAL AND MAIN LINEAMENT MAP
OF BASTAR BLOCK, M. P.
(BASED ON INTERPRETATION OF LANDSAT MSS IMAGERY)
SCALE
0 20 40 Kms.
LEGEND
12 ALLUVIUM
Recent
11 UPPER GONDWANA GROUP
Middle Triassic to Lower Cretaceous
10 LOWER GONDWANA GROUP
Upper Carboniferous to Lower Triassic
9 CHHATTISGARH GROUP INDRAVATI GROUP SULLAVAI GROUP PAKHAL GROUP
8 NARAINPUR GROUP
UPPER PROTEROZOIC
7 GABBRO ANORTHOSITE COMPLEX
6 DONGARGARH GRANITE
5 NANDGAON GROUP
MIDDLE PROTEROZOIC
4 BAILADILA GROUP
3 BENGPAL GROUP
LOWER PROTEROZOIC
2 KHONDALITE / CHARNOCKITE GROUP
1 UNCLASSIFIED GNEISSES
ARCHAEAN
MAIN LINEAMENTS
Lohara
Dhamtari
Binda Nawagarh
Bolangir
Mohala
Kanker
Bhanupratpur
Keskal
Muramgaon
Antagarh
Bhawanipatna
Pratabpur
Narayanpur
Kondagaon
Surajgarh
Dringi
Bhamragarh
INDRAVATI
Jagdalpur
Jeypore
Koraput
Darba
Mandavgi
Bhopalpatanam
Samasgarh
MAHANADI
KOTRI R
INDRAVATI R
GODAVARI R
V. D. Bhate.

Figure – 1

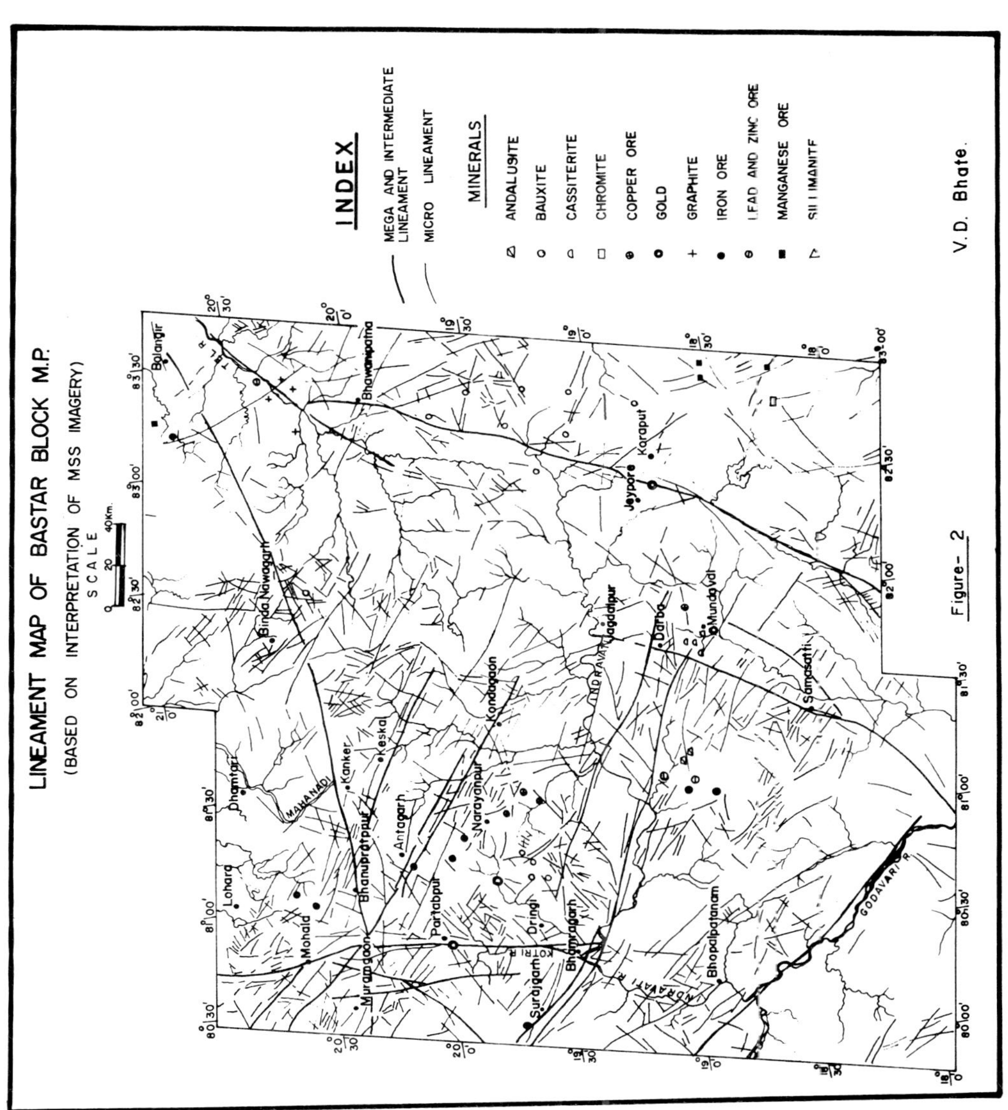
LINEAMENT MAP OF BASTAR BLOCK M.P.
(BASED ON INTERPRETATION OF MSS IMAGERY)
SCALE
0 20 40Km.
INDEX
MEGA AND INTERMEDIATE LINEAMENT
MICRO LINEAMENT
MINERALS
ANDALUSITE
BAUXITE
CASSITERITE
CHROMITE
COPPER ORE
GOLD
GRAPHITE
IRON ORE
LEAD AND ZINC ORE
MANGANESE ORE
SILLIMANITE
V.D. Bhate.
Bolangir
Bhawanipatna
Binda Nawagarh
Jeypore
Karaput
Jagdalpur
Darba
Mundaval
Samasatti
Kondagaon
Narayanpur
Keskal
Kanker
Dhamtari
Antagarh
Bhanupratpur
Lohara
Mohala
Murumgaon
Partabpur
Dringi
Surajgarh
Bhamragarh
Bhopalpatanam
MAHANADI
INDRAVATI R
KOTRI R
GODAVARI R
TEL R

Figure- 2

TECTONICS AND LINEAMENT PATTERNS OF THE VINDHYAN BASIN BASED ON LANDSAT IMAGERY DATA

M. Krishnamurty and V. C. Srivastava

Geological Survey of India, Lucknow, India

ABSTRACT

A study of Landsat imagery of the Vindhyan basin covered 124,000 sq. km. on 1:1 million scale extending between the coordinates 24°N to 26°N and 78°E to 84°E was conducted to review the regional tectonics of the basin. The distinct set of lineaments trending (ENE-WSW) along the Son-Narmada and Asmara directions appears to have been responsible in pre-Bijawar times for carving out the Bijawar/Vindhyan basin. There appear to be a series of movements along these geosutures during Bijawar and Semri times and also at the end of Semri time. There are diamondiferous plugs along these two fractures as well as basemetal and radioactive mineralization, thus establishing the idea that they are of deep seated crustal origin.

INTRODUCTION

Landsat imagery of the Vindhyan basin was studied with a view to understanding the regional tectonic setting of the basin. Certain significant tectonic elements have been recognized which appear to have influenced the history of the Vindhyan sedimentation as well as emplacement of diamondiferous plugs and sporadic occurrence of base metal mineralization in the area. The present paper gives a brief account of the nature of the tectonic elements, their influence on the Vindhyan sedimentation and accompanied mineralization in the area.

REGIONAL TECTONIC ELEMENTS

The Vindhyan sedimentary rocks in this area are bound along the south & north by two major deep seated lineaments. The southern lineament herein designated as the Son-Narmada lineaments presents a very wide sheared zone with a number of recognizable major fractures, whereas the northern lineament known as the 'Asmara lineament' appears as a narrow zone and shows comparatively poor surface reflections.

Southern Lineament or the Son-Narmada Lineament. The southern lineament is in fact a composite set of deep seated parallel to subparallel faults traced so far, from Shahabad in the east to north of Katni in the west covering a distance of about 350 km. and is prominently displayed on Landsat imagery. Ground surveys of this tectonic feature revealed that this fault zone separates the Dudhi gneisses and granites in the south from a thick pile of volcano-sedimentary rocks of the Bijawar group in the north. The magnetic survey over the Budhi fault has confirmed its faulted nature between the metasediments in the north and the Dudhi granitoids in the south. The low order S.P.(-50 to -150 m. volt) corroborated by I.P. anomaly of 7.5% maximum P.F.E. against a background of 1%, does not, however indicate any workable base metal deposit. This fault is succeeded in the north by another sympathetic fault named Obra-Amsi-Jiawan fault within the Bijawar group of rocks demarcating the two major and distinct lithological variations of the group. Further north of this fault a number of sympathetic faults are also traceable on the Landsat imagery, the most prominent of which are the Jungel fault and the Son-Narmada fault (locally known as Jamual-Markundi fault). These faults occupy a zone of maximum shear. It is evident from the pattern of faulting that the southernmost Dudhi fault and the Son-Narmada fault of this zone are the two major lineaments which have shaped the present configuration of the Bijawar sedimentary basin.

The northernmost fault of this zone (the Son-Narmada lineament) limits the southernmost boundary of the Vindhyan basin. The Vindhyan-Bijawar contact is marked by a prominent geomorphological break, showing an abutting relationship of Semri group of rocks with the rocks of Bijawar group and sporadic presence of base metal mineralization in Sidhi district, M.P. The Bijawar-Semri contact in the Mirzapur and Sidhi districts of Uttar Pradesh and Madya Pradesh is also marked by the presence of conglomerate which along the major part of the contact is absent as a result of faulting. The reactivation of Son-Narmada fault during the Semri times is further evidenced by the presence of steeply dipping and tightly folded nature of Porcellanite and Kajarahat formations of the Semri group. The flat disposition of upper Vindhyan in the area presents a sharp contrast with the rocks of the lower Vindhyan group which is highly folded and faulted indicating that the reactivation of the Son-Narmada fault died out towards the close of Semri times.

Northern Lineament or the Asmara Lineament. Landsat imagery of the northern part of the Vindhyan syncline suggests the presence of an equally extensive major lineament separating the Bundelkhand granite massif on the north from the Bijawar and Vindhyan rocks on the south as in Sonrai area. This fault also like the southern fault constitutes a composite suite of sympathetic fractures herein named as the 'Asmara Lineament'. Although the Asmara lineament has been identified in aerial photographs and Landsat imagery extending within the upper Vindhyan (Kaimur plateau), it is strongly suggestive that the major component of this system of faults is actually buried under the Gangetic Alluvium. The Bijawar formations in this area are folded in the form of a WSW plunging synclinorium. The northern limb of this fold is traversed by several strike and oblique faults and lies with a thrust contact along the Bundelkhand granite massif. Although the extent of this major fault separating the Bijawars from the crystallines is not fully traceable in the area, it is reasonably believed that it is coextensive in nature with the Asmara lineament to its south. Base metal and uranium mineralization in the sheared zone of Bijawar group of rocks as well as the kimberlite plugs at Majhgawan and Inota along the Asmara lineament indicate the presence of a fundamental fault along the northern boundary of the Vindhyan basin. Geomorphologically the Vindhyan basin presents a crescent shape having length and breadth ratio of 5:1. This remarkably linear tract of sedimentary rock has, therefore, developed as a result of trough faulting due to the activation along the two deep seated major lineaments, namely the Dudhi/Son-Narmada lineament and the Asmara lineament.

The tectonic history of the area, as it emerges from the present study indicates that major geosutures corresponding to ENE-WSW trending Dudhi/Son-Narmada and Asmara directions were developed in this part of the peninsular shield during the pre-Bijawar times, resulting in the formation of a linear trough which received sediments during Bijawar and Vindhyan times. During and after the deposition of Bijawar rocks both the southern and northern geosutures were reactivated as a result of which a number of sympathetic faults with attendant cross faults developed along the margin of the basin. The Son-Narmada fault was specifically activated to a greater magnitude and the movement along this fault was mainly responsible for the development of the ultimate Vindhyan basin.

Trend lines observed on Landsat images and aerial photographs along the southern margin of the basin appear to be the result of translatory movements along ENE-WSW direction particularly along the Son-Narmada fault. A comparatively wider aerial extent of Bijawar rocks along the Rihand river section is the result of this movement. As a result of reactivation along Dudhi and Son-Narmada faults the Bijawar terrain was uplifted and restricted deposition of Semri deposits towards the north. The folded and faulted nature of the Semri group suggests that the Son-Narmada fault remained tectonically active up to the close of Semri time.

The geosuture along the northern margin of the basin likewise developed during pre-Bijawar time and demarcated the northern boundary of the Bijawar sediments. This fault was reactivated during as well as after the deposition of the Bijawar formation. The Asmara lineament existed as a deep seated fracture sympathetic to the major fault along the crystalline margin. This fault seems to have remained tectonically inactive during the deposition of Kaimur rocks. The present day surface reflections over this fault are the result of the fracture in the Kaimur basement.

MINERALIZATION

The intense tectonic activity along the margins as well as in the central parts of the Bijawar-Vindhyan basin was accompanied by silicification of the country rocks, intrusions of quartz veins and basic and ultrabasics along the major fault zones. The mineralizing solutions similarly found suitable pathways along these fault zones.

Silicification of the country rocks and presence of quartz reefs along the Dudhi fault in the south are the instances of potential mineralized zones in this area. Chalcopyrite and galena disseminations as well as ubiquitous presence of pyrite has been noticed in these quartz reefs, especially in the Labhri area. Basic metavolcanics and amphibolites are also found as intrusive within the Dudhi fault zone.

The Jungel valley fault within the Bijawar terrain is marked by the presence of a complex assemblage of basal agglomerates, tuffs and kimberlites. Recent exploration in this zone revealed the presence of a number of kimberlite plugs aligned in a ENE-WSW trend. This trend is also confirmed by the magnetic iso-anomaly contour map. Residual gravity closure in the Jungel East area indicates probable deep seated fault controlling the emplacement of kimberlites. These data suggest that the Bijawar experienced several phases of folding, faulting and fracturing and, at places, buckling with crustal openings which formed the loci for volcanism. These openings, some of which could have been abysmal were the channels through which mantle material/kimberlite magma was pushed up in different phases. It would, therefore, reasonably be believed that the extensions of the Jungel fault and several other sympathetic faults and fractures in the area could be potential mineralized zones especially for the location of kimberlite rocks. The northern margin of the Vindhyan basin where a set of faults along the Asmara lineaments have been identified, likewise, bear the evidences of silicification, ferruginization, intense brecciation and basic inclusions. Several roughly circular deep depressions like Rukma, Bairahna and Manikpur-East have been located along this fracture. These depressions bear a striking geomorphic resemblance to the Inota and Majhgawan kimberlite pipes. It is also significant to note that the Inota and Majhgawan kimberlite bodies lie on the WSW continuation of the Asmara lineament. This indicates the extensive nature of this lineament and the attendant complex history of deep seated intrusions. Copper-Uranium mineralization in the Sonrai area which is confined to the brecciated zone, suggests emplacement of mineralizing solutions along this lineament. There are also evidences of considerable hydrothermal alterations of the Bijawar group of rocks.

The entire assemblage of faults representing the two major lineaments along the margins of the Bijawar/Vindhyan basin seem to have a long and complex tectonic history. Evidently the entire marginal areas covered by these deep seated fractures with observed mineralization were genetically connected with the mantle and are the most promising for the search of major mineral deposits.

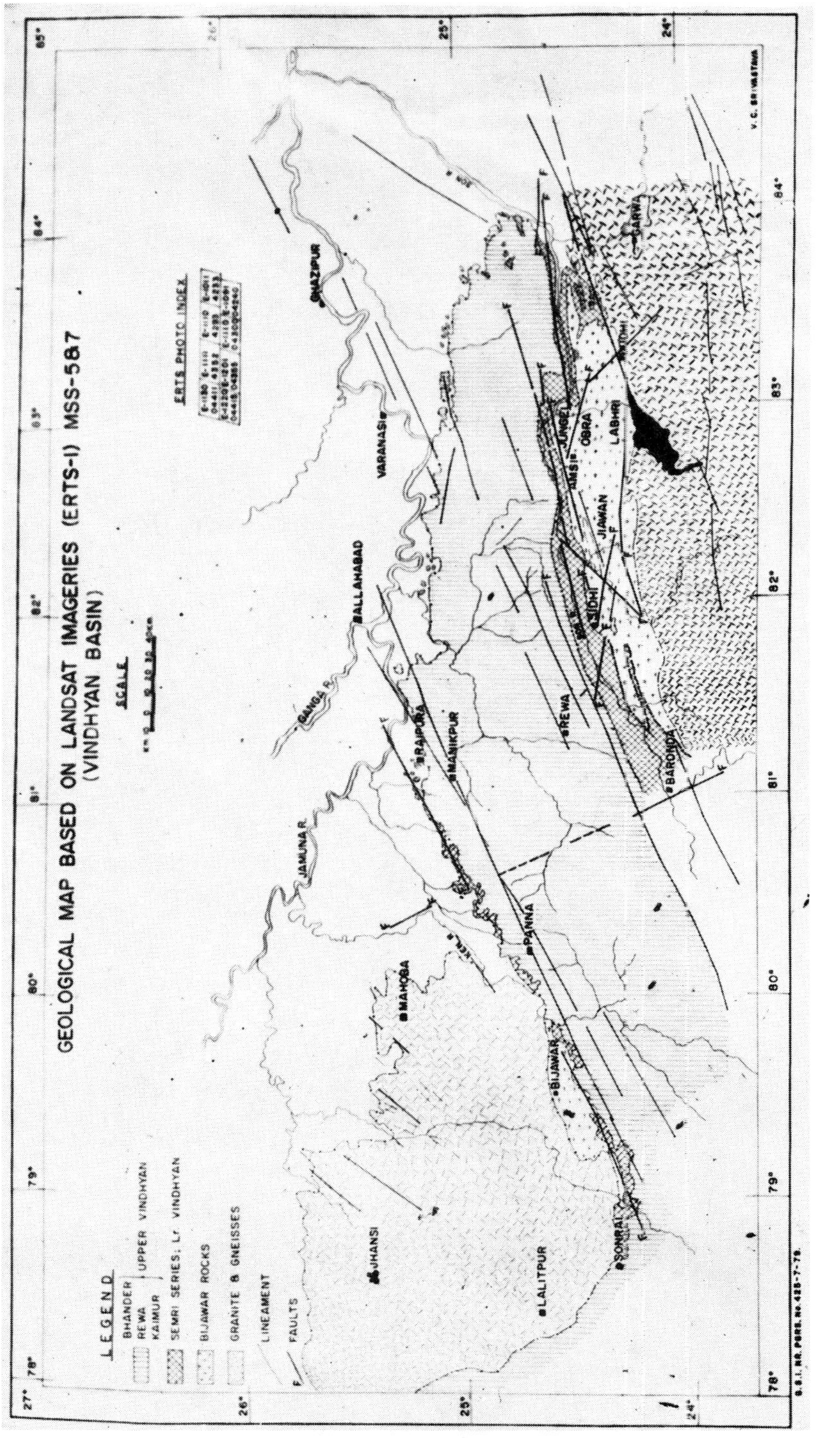
GEOLOGICAL MAP BASED ON LANDSAT IMAGERIES (ERTS-I) MSS-587
(VINDHYAN BASIN)
SCALE
ERTS PHOTO INDEX
LEGEND
BHANDER
REWA
KAIMUR
UPPER VINDHYAN
SEMRI SERIES; Lr. VINDHYAN
BIJAWAR ROCKS
GRANITE & GNEISSES
LINEAMENT
FAULTS
GHAZIPUR
VARANASI
ALLAHABAD
GANGA R.
JAMUNA R.
RAJPURA
MANIKPUR
REWA
SIDHI
JIAWAN
OBRA
LABHRI
BARONDA
PANNA
MAHOBA
BIJAWAR
JHANSI
LALITPUR
SONRAI
V. C. SRIVASTAVA

LINEAMENTS AND THEIR TECTONIC SIGNIFICANCE IN RELATION TO MINERAL POTENTIAL IN SOUTH INDIA

M. Ahmad

Geological Survey of India, Hyderabad, India

ABSTRACT

A lineament map of Peninsular India south of 21°N latitude was prepared involving 51 frames of Landsat I and Landsat II on scale 1:1 million. A study of this map has shown significant relationship of the lineaments in the tectonic history and sedimentation in parts of south India and the preferential location of features like mineralisation, seismicity, hot springs, localisation and movement of groundwater along the lineaments.

INTRODUCTION

A lineament map was prepared using 51 frames of Landsat I and Landsat II covering the Peninsular India south of 21° N latitude. In order to understand their significance, the lineaments were set in a background of the geological map showing location of igneous intrusives, known mineral occurrences of hydrothermal origin, hot springs and epicentres of recorded earthquakes. The lineaments were classified depending upon their lengths as mega (over 300 kms), intermediate (over 100 kms) and minor (less than 100 kms). Mega-lineaments in particular are significant. They cut across different geological units dating from Archaean to Recent and appear to be the surface traces of near vertical crustal fractures and faults. The different trends of the lineaments are NW-SE, NNW-SSE, NE-SW, NNE-SSW, E-W and N-S. The lineaments of the northwesterly and northeasterly trends are most prominent since most of the mega and intermediate lineaments follow these trends. These trends either coincide with or are sympathetic to the known major tectonic zones.

BACKGROUND GEOLOGY

The Archaean and Precambrian formations occupy more than half of the area. The rest is covered by Gondwana and later sediments and by the lava flows of the Deccan Trap (Upper Cretaceous-Lower Eocene). The major mountain building activity involving five periods of folding and metamorphism and igneous intrusions ceased in Precambrian time. The Archaean gneisses and Lower Precambrian pelitic rocks have developed certain persistent regional trends in different parts due to this activity. They are NW-SE to NNW-SSE in the Western part, NE-SW along the Eastern part, ENE-WSW in the central part and NW-SE in the East-central part. Block-faulting and epeorogenic movements in post-Cambrian epochs created locales of sedimentation. These processes continued well into the Neogene and perhaps are still active.

DISCUSSION

Lineaments and their relationship to the structural set up of Peninsular India.

It is interesting to note that the trends of the lineaments mentioned earlier correspond to the regional trends developed in the Precambrian rocks. Several block movements have taken place in the Peninsula. The strike directions of all the blocks invariably follow the present regional strike of rock.

The west Coast of India is a clearly defined fault zone bearing a NNW-SSE direction and is marked by the thick pile of Deccan trap (nearly 2000 m) near the coast. The Godavari graben extends along a NW-SE direction. The East coast, particularly between Ongole and Bhubaneshwar, is parallel to the Eastern Ghat strike. The deposition of the coastal Gondwana and younger sediments in the negative areas are all aligned to the regional strike. Many mega- and intermediate-lineaments correspond to known faults in the above tectonic frame work. In fact, the very triangular shape of Peninsular India appears to have evolved as a result of the influence of NNW and NNE to NE lineaments.

The few examples mentioned indicate that most of the lineaments are zones of structural weaknesses. While some of them appear to have attained equilibrium, others are still active as indicated by the Koyna and Bhadrachallam earthquakes within the last decade.

Lineaments and their relationship to the igneous intrusives

The ultramafic intrusives known in India fall into one or the other belt of mountain building. They include rock types such as serpentine, periodite, saxonite, pyroxenite, epidiorite and amphibolite.

In Karnataka, these ultramafic rocks occur along a NNW-SSE belt in ancient Dharwarian rocks passing through Kadur Hassan and Mysore districts and running parallel to the mega-lineament direction. In the Eastern Ghats the ultramafic intrusives bearing chromite at Kondapalle (Andhra Pradesh) and magnesite in Salem (Tamil Nadu) are observed along NE-SW lineaments.

The Elaeolite syenites of Koraput, Kunnavaram, Elagiri and Sivamalai; the Anorthosites of Khammam, Elchur, Chimakurti, Sittampundi and Kadavur; Carbonatites of Sevathur and Kimberlites of Wajrakarur fall along NE-SW lineaments. In Tamil Nadu several NE-SW trending deep faults like those running along Attur and Kottampatti shear zones contain carbonatite complexes, ultrabasic and alkaline intrusives. The ultramafics of Holenarsapur and Nuggehalli seem to be related to the NW-SE lineaments.

Lineaments and their relationship to the mineral deposition of hydrothermal origin.

In Karnataka, the Kalyadi, Kalsapura and Ingladhalu basemetal mineralisation is located along the NNW lineaments. Also several other shows of base metal mineralisation have preferential localisation along these lineaments. Important gold occurrences of Kolar, Hatti, Gadag and Ramagiri are localised along the NNW trending lineaments.

In Andhra Pradesh, basemetal mineralisation at Gani, Agnigundala, Zangamrajupalli and Mailaram is found in the Cuddapah sediments associated with granitic intrusions and are controlled by ENE-WSW lineaments. These lineaments appear to be the surface traces of faults of post lower Cuddapah period and which extend into the Archaean basement.

In Tamil Nadu, the copper mineralisation at Mamandur is located at the intersection of NNW-SSE and ENE-WSW lineaments.

The examples mentioned above indicate that the lineaments have played a significant role in the structural control of mineralisation. All lineaments, however, are not associated with mineralisation. A detailed study of an area is required to first recognise relationship and environment for mineralisation and then concentrate on those lineaments that show the greatest association with known area. In this way, the lineament study helps to narrow down the areas for mineral investigation.

Lineaments and their relationship to water resources, hot springs and seismicity

Drainage analysis indicates the great control lineaments have played in the development of drainage networks. The history of Cauvery river is a typical example. It originates in the western Ghats and flows in an easterly direction. It was once believed to have flowed in a northeasterly direction and debouched into the Bay of Bengal, a little north of Madras. It was captured by headward erosion of other streams along several lineaments trending NNW and E-W and at present it meets the Bay of Bengal several hundred of kilometres south passing through Tiruchirapalli. This final phase of the course itself is along a WSW-ENE trending lineament.

The Geological Survey of India recently made a detailed study of the groundwater conditions in lineament zones, in the drought prone Anantapur district under 'Operation Anantapur' a multidisciplinary project. By ground surveys including geophysical methods, it was proved that along the lineament, a wide zone of weathering is present both at the surface and in depth. It was also found that the longer the lineament the wider and deeper is the zone of weathering. Wells drilled in this region were subjected to pump test. The best well yielded 100,000 litres of water per hour. By any standards this is a highly encouraging point in the field of groundwater exploration in hard rock areas. Thus, it has been observed, at least in a limited way, that the lineament zones are good source of groundwater.

Several hot springs are known in the Indian Peninsula particularly along the westcoast. The alignment of a chain of thermal springs parallel to the Westcoast between Ratnagiri and Bombay, coincide with the dominant NNW-SSE trend of lineaments. The parallelism of the alignment of Westcoast, thermal springs, the seismic zone and the trend of the dominant megalineaments indicate deep NNW-SSE crustal fracturing of this belt which makes it a potential region for geothermal energy.

The epicentres of a majority of the 53 earthquakes recorded in the peninsular India, between 1594 and 1967 fall on the NNW-SSE megalineament between Bulsar and Bengarla. Included in these is the Koyana Earthquake of 1967. From the irregular intensity distribution of the Koyana earthquake, it is thought that the earthquake originated along a deep, steep fault. Along a NW-SE lineament which follows, in part, the Idamalayar river in Kerala three epicentres have been recorded. In Tamil Nadu a number of epicentres recorded, fall on the NW-SE lineament. The Bhadrachallam earthquake of 1969 falls near the NW-SE megalineament. From this it can be inferred that not only the faults along the coast but also those in the central part of the Peninsula have acted as structural planes of weakness generating earthquakes of varying intensity from time to time. The recent earthquakes have also indicated that some faults are still active.

CONCLUSIONS

1. The present study which is of a nature of observations has shown that lineaments have had a great part to play in the structural and Geomorphological evolution of the Peninsular India, and that the lineaments represent the imprint of the tectonic history of the peninsula in its evolution.

2. Lineaments have played a great part in the control of mineralisation particularly the hydrothermal type, and new areas for prospecting along them can be defined.

3. Lineament analysis greatly help in the exploration for groundwater particularly in the hardrock areas.

4. Correlation of lineaments and the known centres of seismicity helps in the demarcation of seismotectonic zones.

5. Correlation of the lineaments and data on thermal springs helps in indicating areas for exploration of geothermal energy.

INTERFACING WITH SEO TECHNOLOGY: A CASE STUDY IN GEOLOGICAL APPLICATION

J. K. Sircar, S. Chakravarti, P. K. Guha and P. K. Banerji

Photogeology and Remote Sensing Division, Eastern Region, Geological Survey of India, 12 A & B Russel Street, Calcutta 700 071, India

ABSTRACT

The present paper embodies the results of a study on remotely sensed multispectral data, including panchromatic and CIR photographs, from an airborne sensor system with the dual aim of creating the necessary interfacing infrastructure for the utilisation of the anticipated Indian Satellite for Earth Observation (SEO) technology and assessing the utility of data generated in this quasi-operational system simulation of the SEO. Specifically, the study involves an inventory of the interpretational possibilities of the various airborne data products involving both visual and computer aided analysis on a well defined geological problem within a small area of the Bihar Mica Belt, India. The study reveals that visual interpretation based on digitally enhanced data products has a distinct economic advantage over fully automated schemes of pattern recognition.

INTRODUCTION

The near future launch of SEO, the first Indian experimental remote sensing satellite, has made it necessary that the venture be preceded by generation of suitable infrastructure at user levels, so that interfacing with this new technology may result in optimum utilisation of the resultant data.

The work described in this paper is a result of a collaboration program undertaken by the Geological Survey of India, Eastern Region, and Indian Space Research Organisation (ISRO) with the objective of formulating a mechanism for transfer of technology between ISRO and GSI, exemplary in nature for the benefit of other users and at the same time effective for the optimum use of satellite data in geological applications. The geological and tectonic framework of a test area between latitudes 24°23'N and 24°37'N and between longtitudes 85°30'E and 85°45'E forming a part of the Bihar Mica Belt, was studied on a regional scale by remote sensing techniques and subsequently the authors have attempted to indicate its bearing on the ultimate task of prospecting for Mica.

DATA ANALYSIS AND RESULTS

The quasi-SEO system simulated pre-SEO phase airborne data generated by ISRO includes aerial photographs, CIR transparencies and multispectral imageries of 4.2 km square frames with a 15 m spatial resolution in five bands : green (0.5 - 0.6 μm), red (0.6 - 0.7 μm), near IR (0.7 - 0.8 μm), far IR (0.8 - 1.1 μm) and thermal IR (8.5 - 12.5 μm).

A photogeological base map was prepared with limited ground check to provide a basis for further detailed analysis by remote sensing means.

A general guideline for data utilisation and technology transfer mechanism in the field of remote sensing from ISRO to user has been identified by ISRO (1). Accordingly, the authors initially selected a set of frames from ISRO-multispectral data for which ground truth in some details is available. With the aid of an 'Image Processing and Analysis System' developed by ISRO, the investigators singled out the data products that were the most effective. The new information, thus extracted, formed the basis of subsequent interpretation, analysis and development of specific software at user end for further interaction.

The modes of digital processing were outlined by a consideration of the following factors: a) the need to familiarise the user with the techniques of machine-aided processing; b) conceptualisation of the various factors that effect a scheme of processing based on the users need; c) the need to extract data products suitable and relevant to the context of the project and d) the need to enable the user to suitably tailor and design software according to specific requirements around user-end system.

The data base used for digital processing was the formatted CCT, duly pre-processed, containing six scenes of 280 scan lines by 280 pixels each in the five ISRO-multispectral bands in any one file. Once the perspective of the problem and its related objectives are delineated, it is easy to outline a list of requirements for data products generated from the multispectral data. The resultant interpretations are merged finally to prepare a mosaic of the geological themes as required. The initial information extracted from the data set was the distribution of grey levels in a number of selected windows, encompassing all the lithologic variants in the area, in the form of histograms on a line printer. However, a study of the grey tone prints, depicting the pixel by pixel variation in tone in as many as sixteen classes, and histograms suggests a wide variation in surface reflectance characteristics over the same geological formation.

In the context of the analysis, being mainly the classification of different rock types and identification of their differences, it was experienced that the fully automated classification routines, based on statistical pattern recognition algorithms, lead to interpretational difficulties probably due to the lack of representativeness of the training sets - an effect of the inhomogeneity of litho-units presence of gradational contacts, and influence of several types of cover, e.g., vegetation, soil. In the present case, more stress was laid on producing better quality imageries by adopting computer aided enhancement procedures whose visual and interactive interpretation has been by far the most economical and effective in extracting the required information. Contrast stretching in different bands and band ratioed imageries proved to be the most desired form of data product. Both interband relationships and tone - texture characteristics are magnified to a great extent and viewing these enhanced imageries in false colour through an additive colour viewer adds significantly to the possibilities of visual interpretation.

Supplemented with selective field check, the studies reveal (figs. 1 & 2) that in an overall framework of Precambrian metasediments intruded by granites, the plutonic intrusives can be classified into two distinct types : a generally non-porphyritic foliated granite and a typical augen gneiss. The former type of granite is generally associated with a zone of migmatite along the margins. A study of the spatial distribution of pegmatite bodies and their association with respect to the intrusives indicates a concentration of mica-bearing pegmatites around the non-prophyritic granite rather than the augen gneiss.

The interpretation of CIR transparencies revealed that linear features, particularly fractures and faults, are much better depicted in comparison to conventional

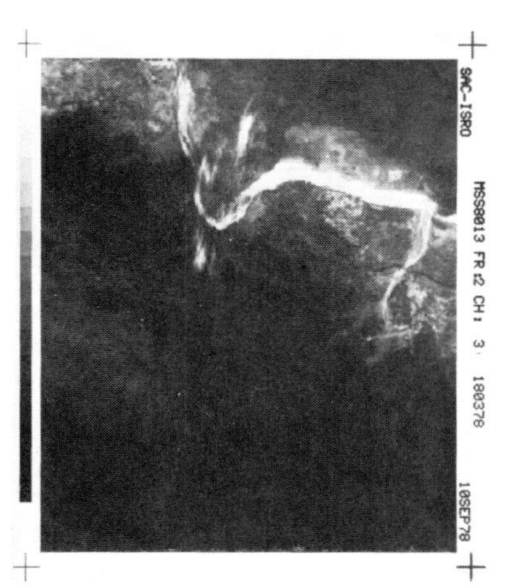

Fig. 1 (left) Original imagery in the 0.7-0.8 m region; (right) Band-ratioed imagery (0.7-0.8 μm/0.8-1.1 μm) (By courtesy of ISRO)

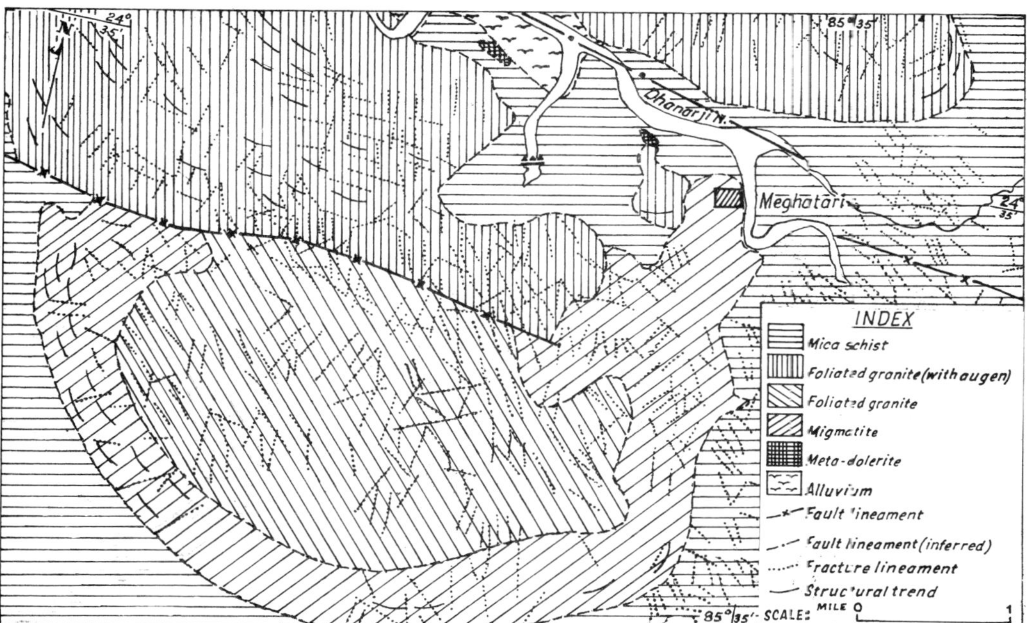

Fig. 2 Geological map of a portion of the project area interpreted from ISRO-Pre SEO data and selective ground check.

black-and-white photographs and provide a distinct advantage in the mapping of linears, an experience shared with others.

CONCLUSIONS

In course of this study, a consideration of the various modes of image analysis reveals a greater utility of an interactive interpretation relying on the visual 'recognition elements', e.g., tone, color, texture, pattern, shape, size etc. on computer aided enhanced forms of multispectral imagery. While it is difficult to generalise a feature selection scheme, it may nevertheless be stated that a distinct economic advantage may be gained by this approach over a fully automated scheme, particularly for large area coverage with limited ground truth.

An analysis of the pre-SEO data products indicates that no single image type is adequate for identifying or mapping of all the geological features. It has been the experience of the authors that emphasis must be laid on what new information can be gained from a particular band, not necessarily which imagery provides the maximum information.

The exercise has resulted in a successful transfer of technology in the field of image processing and analysis. Finally, a few sectors have been identified where detailed investigation for mica may be taken up.

ACKNOWLEDGEMENT

The authors are indebted to the Director General, GSI and the Co-Chairman, Space Application Centre, ISRO, for initiating and supporting the collaboration program and also for their kind permission to reproduce the results at the XXII session of the COSPAR. The authors would like to thank Mr. S. C. Chakravarty, the erstwhile Deputy Director General of GSI, Eastern Region for the encouragement that he had provided to the project.

Acknowledgement is due to all the Scientists at Space Application Centre, ISRO, for their contributions and active participation in the collaboration.

REFERENCE

1) D.S.Kamat, On transfer of Technology with particular reference to digital processing software for remotely sensed data, SEO-SAC-PE 06-79-09-06-01, 1978.

SURFACE CHARACTERISTICS OF PRECAMBRIAN STROMATOLITIC PHOSPHORITES OF A PART OF THE INDIAN SHIELD

D. M. Banerjee

Department of Geology, University of Delhi, Delhi 110007, India

ABSTRACT

The utility of aerial photographs and MSS imagery for exploration of Precambrian phosphorites in a part of Indian shield is discussed. In view of small dimensions of each deposit, the low altitude aerial photography has been found suitable for exploration purpose. LANDSAT MSS images with advance sensors may be found useful in geo-chemical mapping and for identifying individual phosphate units.

INTRODUCTION

In a developing country like India where population is ever increasing despite several measures, increased food production has become necessary to maintain the existing per capita food consumption. This need has stimulated increased resource inputs of many kinds to the food production system to which fertilizers are cogs in the wheel. Phosphorites constitute an important raw material for phosphatic fertilizer. With the present level of mine production of rock phosphate, only 30% of total requirements (for 1978-79) of the country could be met. By 1982-83 the national requirement will be 3.3 m.tonnes against the anticipated indigenous production of only one million tonnes (1). The rest will have to be imported which no doubt is an expensive exercise and requires payment in hard currency or some sort of foreign assistance. Both conditions serve as severe limitations to imports, forcing the farmers to limit their fertilizer usage much below the optimum requirements. Phosphatic fertilizers are in high demand in countries like India, where population pressure requires maximum utilization of agricultural lands. Phosphate removal by crop alone requires the present day production of the world to be doubled. Therefore, fertilizer minerals are in constant need of replenishment, which unfortunately are non-renewable. In order to augment the national and international requirement of phosphatic fertilizer minerals, besides many other minerals like potash etc, new techniques may have to be applied and tested for exploring new deposits in both known and unknown geological formations.

Remote sensing interpretations of LANDSAT Multispectral Scanner (MSS) Imagery has been successfully used in the North Africa and Middle East phosphorite deposits for better understanding of the regional structures, stratigraphic sequences with follow-on low-altitude aerial-gamma spectrometric surveys to locate new phosphate deposits by emitted radioactivity for their contained uranium (2). In addition, new knowledge gained in the recent past about the origin of marine phosphorites may make it possible to identify,'phosphate-loving' lithologies and differentiate them from the 'barren' sequences.

It has been realized that satellite remote sensing may not provide direct clues for pinpointing the phosphorite ore, but it will give us better knowledge of the geology of the deposit. With measurable characteristics of known phosphate deposits on the MSS imagery and by developing new techniques for their analysis, it may be possible to receive information which may help in extrapolating the extension of mineralized areas to other areas besides providing data to develop concepts about regional and local phosphate depositional patterns.

Nature of problem. During the late sixties, when the stromatolitic phosphorite deposits of Rajasthan, a part of the Indian shield, were discovered in thePrecambrian Aravalli Group of metasediments, Geological Survey of India initiated a geological mapping programme in the phosphorite bearing areas of Udaipur and adjacent districts, using 1:30,000 aerial photographs as base maps replacing the conventional topographic sheets. Laboratory based photo-interpretation for phosphorite exploration was not attempted by the Geological Survey, although some preliminary attempts have been made by the author to correlate various phosphorite deposits of Udaipur district on aerial photographs with limited success. Due to other priorities in the area, no serious efforts have been made by any institution or person to establish or negate the utility of aerial photographs in identifying the phosphate beds in hitherto considered 'barren' terrain. This is primarily due to lack of very prominent phosphate features on the available aerial photographs taken on normal panchromatic films.

After the revival of interest in the phosphorite deposits in this country, under the IGCP-156 Project, the entire spectrum of phosphate geology is now being scanned by a group of geoscientists from the Universities, Geological Surveys and exploitation agencies. The present paper is primarily aimed at generating interest in the geological community to undertake serious study of 'phosphogenic' texture and tones on various scales of aerial photographs and on the MSS images. LANDSAT images provide the capability for making better regional maps and defining structures, although it is realized that these capabilities as a guide to exploration is limited by the scale of LANDSAT products. The planners and decision makers of the country are burdened with the responsibility for careful selection and ordering flight missions with narrow band sensors which are most suitable for geologic and earth resources evaluation.

GENERAL GEOLOGY OF PRECAMBRIAN PHOSPHORITES

Most Precambrian phosphorites in India are unique for their characteristic association of algal stromatolitic structures (3). However, exceptions have recently been recorded from the Upper Proterozoic Hirapur Phosphorite Deposit of Madhya Pradesh, where no such algal structures have been found. The major phosphorite bearing lithologies of India are confined to the Aravalli suite of carbonate rocks. The absolute age of Aravalli Group is controversial due to paucity of authentic radiometric data. Crawford (4) has suggested an age of more than 2500 m.y. on the basis of three dates of granites 'supposed' to show intrusive relationship with the surrounding Aravalli metasediments. Sarkar (5) has suggested 950-1500 m.y. for the Aravalli orogeny and metamorphic cycles, thereby suggesting Aravalli sedimentation at a date prior to this. He has however (6) changed his stand and suggested 1600-2500 m.y.for the Aravalli sedimentation. It is now being realized that the geological relationships of various lithostratigraphic units in the area is far from simple and is not yet correctly understood. Consequently, the present author feels that existing radio-metric data need to be interpreted by focussing the geological setting in proper perspective. Based on stromatolitic assemblages and their biostratigraphic correlation with well known and dated sequences in other parts of India, USSR, Australia and China, the Aravalli Group has been considered broadly equivalent to Ripheans, with the oldest element being in the vicinity of 1600 m.y. and the youngest at 950 m.y.(7). These inferences are in contradiction to absolute dates of the so called 'intrusive granites' and therefore leaves a broad scope for fresh radiometric dating of the phosphogenic areas of Aravalli Group of rocks.

Under the present state of knowledge, stromatolitic phosphorites are confined to 'type' Aravalli metasediments of Udaipur and Banswara, 'Aravalli like', but younger to 'type' Aravalli sediments of Jhabua, and homotaxially equivalent dolomitic rocks of Almora-Pithoragarh areas of Kumaun Himalaya. Small occurrences of Precambrian stromatolitic phosphorites have been reported from Shali dolomites of Himachal Pradesh, Vindhyan carbonates of Uttar Pradesh and Madhya Pradesh, Vempalle limestone of Cuddapah and Raipur limestone of Nandini-Drug area in Madhya Pradesh. The Mussoorie phosphorites in Garhwal Himalaya occurring in the lower parts of Tal Formation has yielded cherty phosphatic stromatolites which show close affinity to Precambrian stromatolites (8). This view has recently been substantiated by others favouring a Precambrian age for the Lower Tal Formation in contradiction to commonly held view of Mesozoic-Palaeozoic age for these rocks. Of these,only the deposits of Udaipur and Jhabua have economic viability as phosphorite producing areas due to considerable lateral and depth extensions of the phosphate bearing sediments.

Aravalli phosphorites. Phosphorites in Aravalli 'type' area occur in two distinct forms:

(a) Dark grey phosphorite confined to erect, vertical to branching biohermal colonial stromatolitic columns with well marked phosphate deficient dolomitic intercolumnar spaces. Phosphatic columns stand out in relief due to differential leaching of carbonates and phosphates.

(b) Broken, stromatolitic phosphatic fragments believed to be product of reworking, cemented either by dolomite, chert or clayey material, includes lenticular, bedded, massive phosphates with or without stromatolitic columns.

Due to diagenetic and post-depositional leaching of non-phosphatic carbonates, subsequent collapse of phosphorite fragments and their recementation by silicic or carbonate rich circulating groundwater, the type (b) phosphorite show high

P_2O_5 content per tonne of ore. Three subvarieties can also be distinguished in the field but are distinctly mappable only on moderate scales.

Phosphorites in the comparatively younger sequence of Jhabua show broad similarities with the mappable units of the 'type' Aravallis. A third variety showing distinct creamish-white hue, clearcut bedding character and without any apparent link with algal stromatolites can also be distinguished in a mining block of Jhabua.

Non-stromatolitic phosphorites of the Hirapur-Bassia area of Madhya Pradesh, belonging to Bijawar Group of rocks occur disseminated in highly ferruginous shale-sandstone-dolomitic lithologies and due to high iron-content, stand out in relief as distinct ridges bordering the peneplained Archean granitic complex of Bundelkhand area.

Precambrian stromatolitic phosphorites of Kumaun Himalaya are distinctly associated with crystalline magnesite and dolomite and the phosphorite is either confined to intercolumnar spaces or to the stromatolitic laminae commonly alternating with crystalline magnesite.

The Mussoorie phosphorites of controversial age occur on the two flanks of a major synformal structure in the Garhwal Himalaya and are distinctly associated with bedded and stromatolitic cherts.

Dimensions. Stromatolitic phosphorite bearing units are not sufficiently large to be identified in available orbital imagery. They are however fairly well resolved in 1:30,000 and low altitude 1:15,000 scale aerial photographs. Jhamarkotra deposits of the Udaipur district have the largest dimensions extending for more than 16 km. as a continuous outcrop with 5-50 metres of apparent outcrop width. Deposits at Matoon, Kharbaria-Ka-Gurha, Kanpur, Neemuch Mata etc. are isolated bedded and folded outcrops extending for 1 to 3 km. with 2 to 30 metres of outcrops width. In all these areas surrounding rocks are dolomitic limestone or silicified dolomite . The confining rock units are quartzite and phyllite. In Jhabua, the surrounding rocks are dolomite and granite while the phosphorite beds are 5-30 metres wide and extend as discontinuous lenticular outcrops on the topographic highs for 1 to 5 km. Non-stromatolitic phosphorites of the Hirapur-Bassia area occur on the topographic highs as more or less continuous outcrops for over 5 km. and it is hopefully inferred that this bed joins the phosphorite bed at Lalitpur-Jhansi area, several tens of kilometer away, the intervening area being covered by dense forests and still unscanned in detail. Poorly phosphatic stromatolitic carbonate outcrops of Kumaun Himalaya extend for almost 60 kilometer, covered variably by coniferous vegetation and occurring mostly on the mountain slopes.

Trends. Although structural geometry of individual phosphorite deposits have been worked out, no attempt has been made to relate structural pattern of one deposit with that of another. In the Jhamarkotra deposit, the largest workable phosphorite mine of the country, the phosphorite bed has been folded and refolded to assume the present saucer shaped ENE-WSW configuration, bordering the Archean gneissic cratonic complex. Basement highs on the palaeo-oceanic topography apparently played a decisive role on the phosphate precipitation on the surface bordering the ancient shore-line. The ENE-WSW trend of the central productive zone of Jhamarkotra veers to the NNE-SSW and N-S on flanks of the phosphorite belt, while the

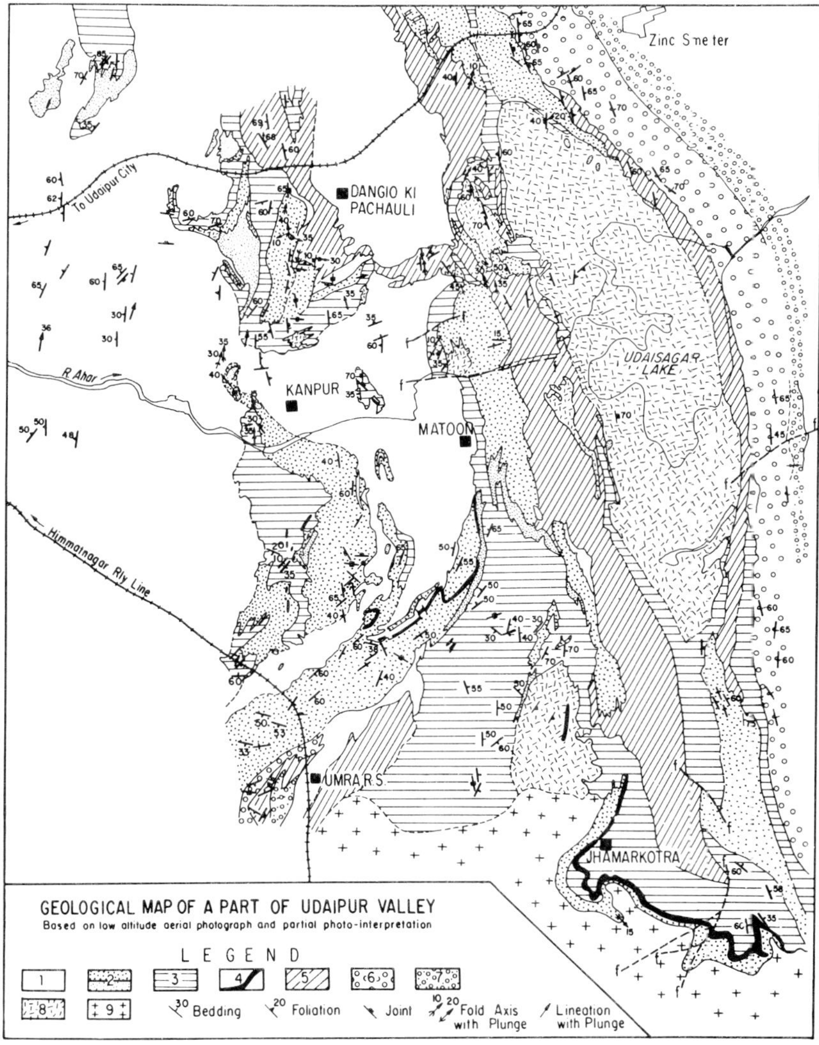

Figure 1. Geological map of Precambrian Aravalli rocks of Udaipur Valley in Rajasthan, reduced from 1:30,000 scale map prepared on aerial photo base with partial photo-interpretation.

Legend: 1. Greywacke-phyllite 2. Orthoquartzite and brecciated cherty quartzite 3. Dolomitic-limestone and other carbonate rocks 4. Phosphorite 5. Carbon phyllite with specks of garnets 6. Basal Aravalli quartzite 7. Basal arkosic conglomerate 8. Granite, possibly intrusive (?) 9. Banded Gneissic Complex

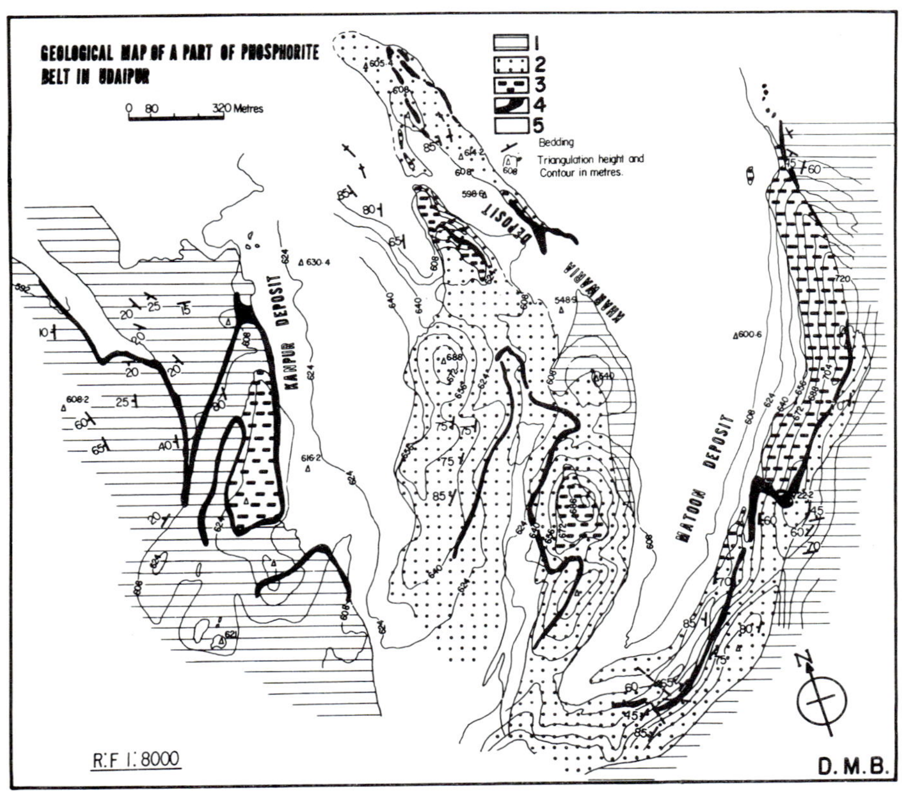

Figure 2. Planetable geological map of three Precambrian Aravalli stromatolitic phosphorite deposits of Udaipur district, Rajasthan. The map is reduced from original prepared on 1:1000 scale. The basic mapping was carried out by several officers of the Geological Survey including the present author.

Legend: 1. Dolomitic limestone 2. Orthoquartzite and quartzite conglomerate
3. Brecciated cherty looking quartzite, and fragmental rocks
4. Phosphorite with algal stromatolites 5. Phyllite, greywacke schist, carbon phyllite

Bedding with dip values — Triangulation height and contour in metres.

Matoon deposit show a broad NNE-SSW arcuate trend. Other smaller deposits of Kharwaria and Kanpur are folded along N-S fold axes, whereas the Neemuch-Mata deposit show variable trends bordering a granitic mass. Jhabua deposits of Madhya Pradesh show distinct N-S lenticularity due to variable N-S plunge of the folds. The phosphorites of Kumaun Himàlaya follow the local E-W Himalayan trend of dolomitic formation. Non-stromatolitic Hirapur-Bassia-Lalitpur phosphorites show a well marked E-W trend. In all cases, phosphorite beds more or less conform to the structural trends of the enclosing host rocks which often show various primary characters of shore-line sedimentation including prolific growth of biohermal tidal to intertidal stromatolites (Fig.1 2 and 3).

Tones and Textures. There is no well marked tonal differences between the phosphorite beds and the surrounding host rock. A careful observation of Jhamarkotra-phosphorite bearing beds on 1:15,000 aerial photos indicate minute light grey circular to irregular spots on the stromatolitic beds, which, if enhanced by optical means, could possibly be used as a distinctive marker for stromatolitic outcrops. Similar textures were also observed on aerial photographs of Neemuch-Mata and Kanpur areas. The aerial photos of the Matoon deposit showed dark shadows on the hill slopes exposing the phosphorite bed and consequently could not be studied for tonal and textural relationships.

Using textural-tonal and structural criteria, it was possible to infer that the Jhamarkotra, Matoon and Neemuch Mata phosphorite beds are not one single unit and are separated by intervening dolomite and greywacke-like sediments which further substantiates the author's hypothesis of recurring phosphate-precipitation events within the compressed Aravalli sedimentation history.

As mentioned before, the phosphorite beds are in most cases surrounded by dolomitic lithology, although at times these occur on the fringe of the carbonate-host rock either bordering the granitic craton or remaining bounded by shallow marine quartz-arenites. In such cases, phosphorite-dolomite units show distinct surface-textures compared to granitic-gneisses or quartzites reflected by joint and gully patterns. No such distinction has been possible so far in the orbital imagery.

MSS Images. Being outcrops of limited dimension, the phosphorite beds of Udaipur region have not been identified on LANDSAT multispectral scanner images. The entire metasedimentary unit of Aravalli Supergroup could however be picked up as showing distinct texture and pattern along a linear belt extending roughly in N-S to NNE-SSW direction, confined by cratonic masses of Pre-Aravalli (Archean) gneissic basement complex. Imagining the ancient shore-line along the margins of the basement complex and possible presence of sea-mounts in the Aravalli sedimentation basin leading to deposition of shallow tidal-intertidal sediments in an otherwise shelf-sedimentation regime, possible areas of biohermal carbonate sedimentation can be inferred on the MSS imagery. Through these inferences, one can logically suspect possible locales of stromatolitic phosphorites in the Aravalli Group of lithologies. Image analysis should be done at various levels of stratigraphic sequence considering the repetitive nature of phosphorite formation through geological time. Resolution and albedo contrast in other Bands except Band-7 is too poor for deciphering the lithological boundaries. In Bands 6 and 7, tectonic units can be easily discerned, thereby revealing the deep seated fracture systems traversing the area (fig. 4).

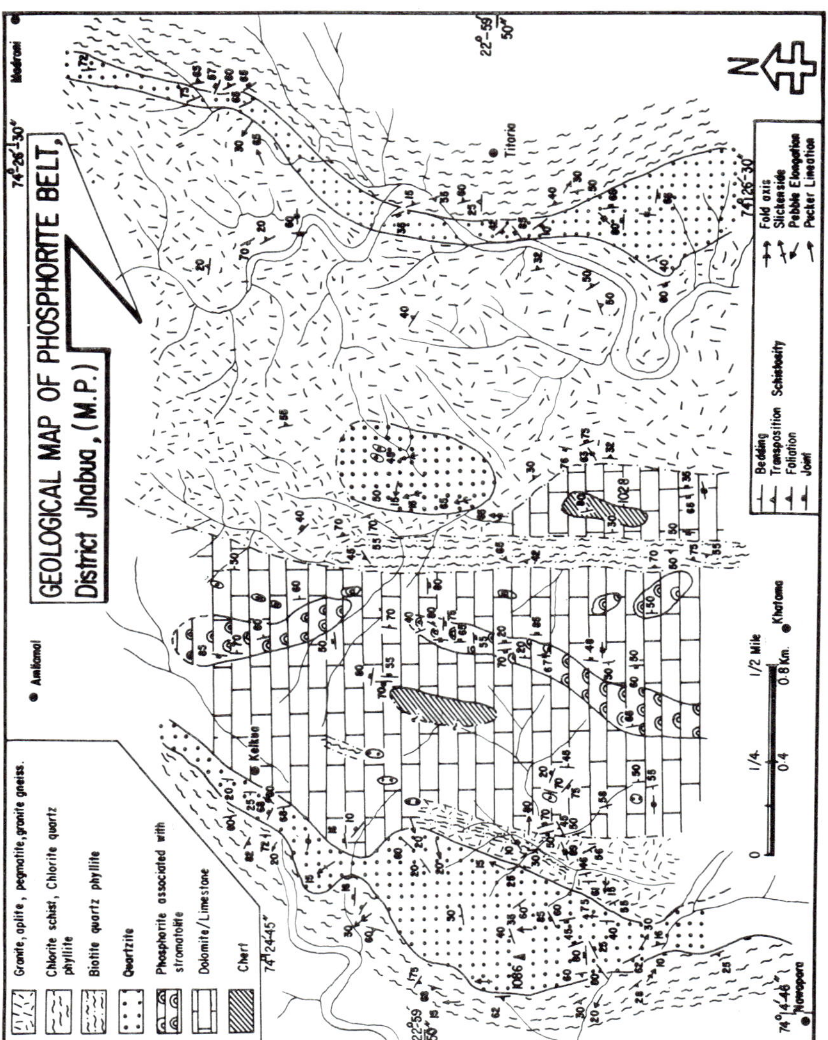

Figure 3. Geological map of Precambrian stromatolitic phosphorites of Jhabua area in Madhya Pradesh. Original mapping was done by the author and P.C. Basu on conventional topographic sheets.

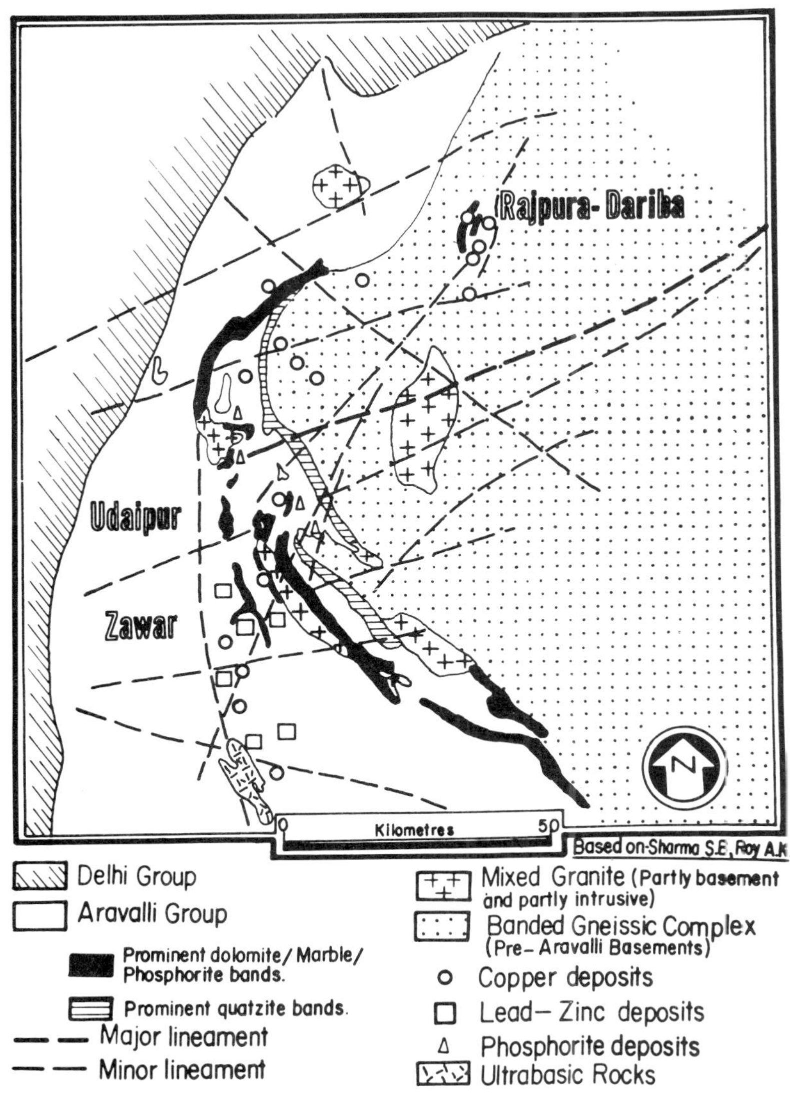

Figure 4. Consolidated lithological and lineament map of phosphorite and base-metal belt of Udaipur region, Rajasthan, based on LANDSAT Imagery. The interpretation is based on works of S.B. Sharma and A.K. Roy of Geological Survey of India.

Except for Udaipur, no other phosphogenic area has been studied either on MSS Imagery or low-altitude aerial photographs. No spectral measurements have so far been made in the field or in the laboratory.

A PROPOSAL FOR FUTURE

Synsedimentary mineral deposits like phosphorites are related to chemically receptive rocks which favoured phosphate mineralization. Such rocks are usually porous, permeable or highly fractured dolomites, limestones, cherts and breccias. With a minimum of ground-truth, the rock types in such areas can be broadly subdivided. This is however not possible in vegetated areas. In addition, topographic expressions of mineralization and alteration represent rare geological events in areas of tropical weathering, where residual ore deposits are related to topography. These minerals concentrate on erosion surfaces, whose identification may serve as an important guide to phosphate deposits.

In phosphogenic areas, rock types can be inferred from known spectral properties. If the rock types are 'ridge makers', as some of the phosphorites are, they can be easily picked up as marker beds. Using LANDSAT imagery, one can identify promising areas of phosphate accumulation, decipher the broad tectonic setting favourable for phosphate trapping and pick up the most promising areas for follow-up aerial photography, field checking and drilling.

Finally, it is suggested that geochemical mapping with the help of analysis of MSS Scanner images should be attempted to map chemical differences among dolomite-phosphorite-quartzite and granite. Geochemical mapping, however, requires accurate radiometric data (the radiance detected in each scanner channel) which necessitates corrections for spurious environmental effects, such as spatial and temporal changes in solar illumination and atmospheric parameters (9). In order to make use of remote sensing data for geochemical mapping, the geological targets should possess diagnostic emittance and reflectance features correlatable to their chemical composition. It has been indicated (10) that the spectral emittance features in the thermal infra-red regions are caused principally by inter-atomic motions in silicate, carbonate and phosphate minerals. Mapping by spectral reflectance and emittance features of various minerals has been discussed by Vincent (10), wherein he has emphasized the importance of spectral ratioing. Spectral emittance data in 8-14μm spectral region is likely to give important compositional information of phosphate minerals in particular. As postulated by workers in the field of space-geology, the author believes that future scanners with multiple channels in 8-14μm wavelength regions along with visible and reflective infra-red channels will be useful for solving many problems of differentiating lithologies on the basis of their chemical differences. Various data enhancement techniques such as colour adaptive viewing, band ratioing and colour photographic compositing should be used with MSS Scanning.

ACKNOWLEDGEMENTS

This paper has been written on receiving a request from W. D. Carter of U.S.G.S. who himself suggested the author the likely contents of the paper. The author is grateful to him for the initiative and the invitation for presenting the paper in the COSPAR Workshop. ISRO-SAC-UGC-INSA have jointly shared the cost for author's participation in the Workshop and the author is grateful to all these organizations. Sincere thanks are due to Dr. M.V.N. Murthy and Dr. S.B. Sharma of Geological Survey of India (W.R) for allowing the author to scan through aerial photographs and MSS imagery of Udaipur region in their Jaipur Office.

REFERENCES

1. I.G. Jhingran, Subramanian, K.R.V. and Choudhuri, R., Chemical Times, XXII-XXVI (Dec 4, 1978).

2. W.L. Smith, in: Remote Sensing Applications for Mineral Exploration, Dowden-Hutchinson & Ross, Inc. 1977, p.9-27.

3. D.M. Banerjee, IX Congress Int. De Sedimentologie, Nice, France. Theme 9 (1975).

4. A.R. Crawford, Can. Jour. Earth Science, 7, 1, 91-110 (1970).

5. S.N. Sarkar, Dhanbad Publishers, India, 1968, 33 p.

6. S.N. Sarkar, 24th Int. Geol. Congress, Sect. 1, Montreal, (1972), p.260-272.

7. D.M. Banerjee, Jour. Geol. Soc. India, 12, 4, 349-355 (1971).

8. D.M. Banerjee, Workshop on stromatolites: Characteristics & Utility, Udaipur, (Abstract), p.12 (1978).

9. W.L. Smith, in: Remote Sensing Applications for Mineral Exploration, Dowden, Hutchinson & Ross, Inc. (1977), p.73-98.

10. R.K. Vincent, in: Remote Sensing Applications for Mineral Exploration, Dowden, Hutchinson & Ross, Inc. 1977, p.251-277.

APPLICATION OF REMOTE SENSING TECHNIQUES TO PETROLEUM EXPLORATION IN INDIA

S. N. Talukdar

Institute of Petroleum Exploration, Oil and Natural Gas Commission, Dehra Dun, India

ABSTRACT

The extensive use of remote sensing techniques in petroleum exploration in India had a modest beginning in 1930's when the geologists of Assam Oil Company employed photogeological methods to map the densly forested, and highly inaccessible areas in Assam, Tripura and Mizoram in the eastern part of India. After the 'Oil and Natural Gas Commission' was set up in 1956, to explore for petroleum in India, photogeological and photogeomorphological studies were extended to cover all the sedimentary basins of interest in India. With the availability of Landsat data, both visual and machine aided interpretation techniques have greatly facilitated our oil exploration efforts. An outline of the activities of Oil and Natural Gas Commission in this fascinating field is presented in this paper.

INTRODUCTION

The Assam Oil Company was the only organization exploring for petroleum in India until the middle of this century. Their lease area comprised the thickly forested highly inaccessible terrain in Assam, Tripura and Mizoram. With the advent of aerial photography those pioneers realised the value of photogeological interpretation and systematically covered their lease area by such studies. Many prospective structures were thus discovered by them.

When the Oil and Natural Gas Commission (ONGC) was set up in 1956 as a government agency to explore for petroleum in India, one of its first acts was to develop a well equipped photogeological laboratory since it was realised that only by using air photo techniques the vast unexplored areas could be covered with speed and economy. To start with, geologists were trained in the photogeological laboratories of Royal Dutch Shell Oil Company (B.I.P.M.) in The Hague, and in the International Training Centre for Aerial Surveys in Delft, Netherlands and regular productive work started from 1958 onwards. ONGC has not turned back since. By the time the Indian Photo Interpretation Institute was set up in Dehradun in 1965, with the help of I.T.C. and the Dutch government, Oil and Natural Gas Commission had already developed its own expertise and was only too happy to extend a helping hand in the formative stages of that now renowned institute.

PHOTOGEOLOGICAL STUDIES

The initial geological activities of ONGC were centred around the Himalayan foothills extending from Indo-Pakistan border in the west to the Nepal border in the east. The project was to map these Tertiary sedimentary tracts in detail. Maps with preliminary interpretation were checked in the field and the final geological maps contained both photogeological and ground data. The speed and economy of this technique enabled the organization to cover all the sedimentary basins within a shorter period compared to ground surveys alone. As an illustration, the Balh anticline near Jawalamukhi in Kangra District of Himachal Pradesh took me and a party of three geologists 16 man months to map on 8" = 1 mile scale in 1957 and its closed area is only 155 square Kilometres. After the advent of photogeological techniques, the complete 28,000 square kilometre area in Mizoram was mapped in 38 man months including the necessary selective ground checks. ONGC has since completed photogeological mapping of all the exposed sedimentary tracts of India. Presently aerial photographs are being used essentially for solving special problems like studying the effects of neotectonic movements in Godavari-Pranhita valley, planning a pipeline alignment in Tripura or Gujerat or wherever revision has become necessary.

PHOTOGEOMORPHOLOGICAL STUDIES

The Indo-Stanvac project exploring for petroleum in the late fifties had obtained the services of Charles De Bleuix, an American photo-geologist with considerable experience in Mississipi delta area, to study the airphotos of the alluvium covered West Bengal Basin. He recognised quite a few tonal and geomorphological anomalies on aerial photo mosaics and many of them later turned out to be expressions of subsurface structures. A few of us were associated with this study and ONGC took up similar photogeomorphological studies of all the alluvium covered sedimentary basins of India. These studies gave valuable leads in the eastern coastal basins, Cambay Basin and Brahmaputra Valley. The well known Anklesvar and Kalol oilfields of Gujerat are reflected as geomorphic anomalies. In Cauvery Basin along the east coast of India, seventeen localities were identified as anomalies based on air photo study. Subsequent geophysical surveys confirmed most of them as subsurface structures. One such anomaly around Tirukadaiyur to the north of Karaikal which did not show up with conventional seismic surveys came up nicely as a subsurface structure when more sophisticated digital CDP seismic work was carried out. This structure is still a prospect waiting to be tested. Similarly many localities have been identified in different alluvium covered basins which would eventually be covered by more definitive seismic surveys and drilling. Experimental work carried out in the dune areas of Rajasthan has indicated a relationship between the type of sand dunes and the subsurface structural configuration.

LANDSAT STUDIES

With the availability of Landsat data from the middle 70's, an additional tool has been provided to the oil explorationists. These are now very well known, and their utility and benefits have been well advertised. Landsat imagery have their own advantages as well disadvantages, but one thing is apparent, they cannot displace the large scale conventional aerial photographs for detailed mapping purposes. However, we have to use all possible techniques in oil exploration since detection of new structures has to be a continuing endeavour.

The advantages offered by Landsat imagery are well known. The synoptic view obtained, the uniform illumination, repetitive coverage and amenability of data for digital and optico-electronic processing - each one of these characteristics can be profitably utilized in petroleum exploration. Oil companies are the principal users of Landsat data from EROS Data Center and they are carrying out extensive work with them. The exact nature of research and the results obtained are largely kept confidential due to the competitive nature of the oil business. The participation by oil company geologists in relatively large numbers in the recent ERIM sysmposium at Ann Arbor reflects their continued interest. The Geosat Committee of U.S.A. is also dominated by oil companies. In the October, 1977 issue of Aviation Week and Space Technology, it is claimed that Earth Satellite Corporation of USA has developed a billion dollars worth of oil through remote sensing techniques. The main emphasis appears to be on mapping of megalineaments reflecting deep seated discontinuities not all of which are deducible even with the availability of a great deal of subsurface information and very subtle tonal anomalies not otherwise detectable without the flexibility of processing available to digitally recorded data. We are familiar with Dr. W.D. Carter's work on megalineaments in U.S.A. and the occurrence of major oil fields around the huge curvilinear feature situated in the central part of U.S.A. In the recent A.A.P.G. symposium at Houston during April 1979, Dr. Michel T. Halbouty presented a paper in which he stated that all the giant oilfields of the world would have been easily recognised on Landsat imagery had they been available at the time they were being explored. In ONGC, we are carrying out experimental work in the following aspects employing visual interpretation techniques as well as machine aided ones:

) Detection of subtle tonal anomalies which could reflect concealed structures below the alluvial cover of sedimentary basins. The Cambay Basin in Gujerat has been taken up as a test area. Airphoto studies had earlier indicated many anomalies and Landsat studies using digital enhancing techniques are now being conducted to make multistage interpretation of this area. As will be mentioned later, the Space Application Centre of Indian Space Research Organisation is carrying out thermal scanner surveys as a cooperative endeavour with the ONGC in the same area. There are good chances of finding some more anomalies in this region by these means.

) Delineation of signatures over established oil fields and subsequent extrapolation of the same into immediately adjoining unexplored areas. In the Brahmaputra valley of Assam, ONGC's activities are mainly concentrated in the areas to the south of the river and many oilfields have been discovered already. In the area to the north of the river, the available data is scanty. We are extrapolating the oil field signatures to these northern parts and the promising areas are proposed to be covered by detailed geophysical surveys.

3) Systematic comparison of lineament pattern with geophysical data to bring out an integrated interpretation of basin development: A close relationship between the gravity trends and lineament pattern has been observed in many areas and the studies are still in progress.

4) Extrapolation to offshore areas of the major onland structural trends to facilitate better interpretation of offshore data (seismic profiles and bathymetric charts). A detailed analysis of the megalineaments of the Peninsular area of South India has aided the interpretation of the offshore data along Kerala and Tamil Nadu coasts. Similar studies in Andamans have indicated the emergence of an island arc system to the west of the island chain.

ONGC has an exclusive Rs.2.5 million project under the National Council of Science and Technology scheme to develop remote sensing techniques in oil exploration. This project includes developing digital and optico-electronic techniques to identify and enhance oilfield signatures. Under this project, ONGC is shortly acquiring one International Imaging System's Video Image Processor which would be a good stand alone device for quick image enhancement and also would be helpful in predigital sorting and classifying. Our computer and software division is also associated with remote sensing and necessary software is being developed indigenously to process and enhance Landsat data in ONGC's IBM 370/145 computer. We realise that we cannot afford to wait until our own software is perfected, and in order to satisfy our immediate requirements we are looking for collaborators both within and outside our country. To understand the evolution of any basin, all the available data should be integrated. The data from remote sensing should be combined with the available geophysical information if a prospect has to be identified properly. I know that the U.S.G.S. and CSIRO of Australi have developed a technique by which a stereoimage is simulated by computer using either gravity or magnetic data for creating the third dimension in the otherwise two dimensional Landsat imagery. This has a great potential in oil exploration and we are exploring the possibilities of collaborating with U.S.G.S. in this regard. Another important aspect is to get geometrically corrected versions of computer tapes in which Landsat data have been resampled to a 50 metre x 50 metre grid, and in which the 4 channels of Landsat multispectral data have been rotated to a west-east scan line direction so that the resampled tapes can produce images which can be directly superimposed on our standard quarter inch ($1^o \times 1^o$) topographical maps. This is very necessary in order to superimposed directly other geophysical data on the same scale over the Landsat imagery. Geo-spectra Corporation of Ann Arbor has developed the necessary software for this technique and also for digitising terrain data and producing images from them similar to shaded relief maps, so as to enhance structural features which have subtle topographic expression. The product resembles radar imagery with the difference that we can assign the sun position to any angle or direction we choose! This technique has been found to be very useful for oil and gas exploration in Michigan Basin even though Pleistocene glacial tills of 20 meters to 200 meters thickness overlie the Jurassic bed rock. These are interesting possibilities which could hopefully discover for us a few more oil fields.

AIRBORNE LINE SCANNERS

Whether it is airborne or spaceborne, the principle of recording in the multispectral scanner is identical. Of course in the airborne scanner, we have a choice of more number of spectral bands. For geological interpretation, our impression is that spectral information in the near infrared (1.55 to 1.75 um) and thermal infrared (10.4 to 12.5 um) is ideal. However a lot of research and development effort is required to establish the band or combination of bands which could enhance oilfield signatures. I am at present not very sure of the economics of this, mainly because of the high expenditure involved which may not be justifiable if we consider the marginal amount of additional spectral information it could give. The digital processing is also made very costly by the inherent distortions in gathering the airborne scanner data. At present cost-benefit base level is around what the conventional aerial cameras fitted with black and white, color and infrared films provide. However, let me hasten to add that we have not closed our options. We have taken up seriously a cooperative project with the Space Applications Centre at Ahmedabad, in which they have already flown their thermal scanner over a few established oilfields in Gujerat. In my sincere belief, this project should blossom into a long-term, mutually beneficial endeavour. Before long we should be able to establish, jointly, a system designed for oil exploration purposes, which could be incorporated in our own ongoing and future space missions. I am very hopeful about the success of this project. I am also happy that the Space Applications Centre is already on the job in developing charge coupled device detectors which could be used in a 'pushbroom' scanning mode similar to what the French are proposing to incorporate in their SPOT satellite. I wish them all success, since that will remove substantially my apprehension about the cost-benefit level of processing line scanner imagery. They should also give serious consideration to obtaining stereoscopic coverage and increase the resolution to an optimum level. Then we can say that we are on the freeway!

Remote sensing through satellites and airplanes as an aid to oil exploration has come to stay and we should now strive to develop more and more sophisticated techniques at lesser cost, especially because it is going to be more and more difficult for us to take out the remaining bit of oil that still lies undetected in the depths of the earth. We have already mapped all the easily recognisable structures, and it is when we have to delineate the more difficult and more subtle structures, that I turn to remote sensing techniques very hopefully indeed. Up to now, in remote sensing, we have been looking at the irregularities or perturbations which a potential trap in the subsurface gives rise to on the surface topography. We should not stop with this one aspect of satellite sensing capability. The ultimate aim of an oil explorationist is to find a device for the direct detection of hydrocarbons under the ground. With the tremendous progress in information theory and signal processing achieved by man's space efforts, the time is not too far away when remote sensing will enable us to detect minute variations caused by actual presence of oil in the subsurface by remote sensing techniques.

The developing countries are certainly grateful to NASA for making available Landsat data without any restrictions. However, the availability of data from future satellites like magsat, and stereosat to developing nations is not certain. Under the circumstances, I am all the more eagerly looking forward to the Indian earth resources satellites. They may not be the latest in instrument sophistication, but we are proceeding in the right direction.

ACKNOWLEDGEMENT

I would like to express my thanks to the organisers of this workshop for inviting me to present this, the only paper concerning petroleum exploration, and in particular to Dr. W.D. Carter of U.S.G.S. and Mr. D.S. Kamat of SAC. I am grateful to the Chairman, ONGC for permitting me to present this paper.

EVALUATION OF MSS IMAGERY OVER PART OF THE ASBESTOS-BARYTES BELT OF SOUTH-WESTERN CUDDAPAH BASIN, ANDHRA PRADESH, INDIA

D. N. Setti and K. Krishnanunni

Geological Survey of India, Calcutta 700016, India

Aerial photographs have been increasingly utilised in India in geological appraisal. With the advent of Landsat I and the commissioning of multi-spectral airborne scanner by the National Remote Sensing Agency (NRSA) in 1977 data from these has also started being utilised.

The distinct lithological and or structural control of asbestos and barytes in a well banded sedimentary sequence, close to the Archaean basement with a lot of easily identifiable geological features, low relief and consequently low distortion on airborne scanner imagery, prompted the choice of the study area for a comparison of information output and interpretability of the different Landsat imagery bands, medium scale aerial photography and airborne multi-spectral scanner imagery for optimising spectral bands for geological exploration.

Study material   Landsat imagery on 1 : 250,000 scale in all the 4 bands, black and white photography on 1 : 32,000, scale 1 : 63360 topographic sheets and airborne MSS data collected by the 11-channel Bendix Modular Multi-spectral Scanner (M2S) from an altitude of 2600 m. formatted on 70 mm film strips were used. Of the 11 channel M2S imagery three with poor contrast were excluded. The imagery details including spectral band widths are illustrated in Fig. 1. The interpretation was restricted to an area of 360 sq km. covered on 3 M2S lines oriented north-south. This is only a small part of the Landsat frame and is covered by 40 aerial photographs.

Geology and mineralisation   The study area forms a part of the south-western portion of the crescent-shaped Cuddapah basin. Unconformably overlying the Archaean basement of granites and gneisses is the Gulcheru Conglomerate and the Vempalle Formation comprising dolomite, shale and chert with sub-aerial and submarine basic volcanic flows and sills. Over this is the impersistent Pulivendla Quartzite and the Tadpatri Shale with intercalations of limestone, volcanic flows and basic sills. All the beds dip gently towards the basin and are displaced by E-W trending faults.

The asbestos prospects spread over a 16 km long NW-SE trending belt in the Vempalle dolomite and are confined to the serpentinised upper contact (over a metre thick) of the dolomite and the lowermost dolerite sill. The veins are about a centimetre in thickness and are erratic in occurrence. Barytes occurs as veins upto 1 m. thick, in east-west trending faults and fractures in the basic sills of the Vempalle Formation.

Method of study The satellite imagery on 4 bands and the 8 channels of M2S imagery were interpreted singly and as pairwise combinations, of which three are illustrated in Fig. 2,5,6. The M2S imagery has a mean scale of 1 : 80,000 along flight direction (N-S) and 1 : 50,000 across. The thermal band (Channel-11) could not be used as it had too few details. Interpretation was based on recognition of topographic and cultural features, lithological units, structures and features characteristic of mineralisation. Under each of these broad groups, the interpretability of 4 to 6 specific sub-features have been ranked from 0 (not identified) to 5 (easily identified) (Table - 1).

The mineralisation per se is not identified on any of the image types because of extreme small size. However, the presence of dumps and workings aligned along specific directions within distinct host rocks offer clues for extrapolation on air photos and, to a lesser extent, on some M2S imagery.

General Observations : Landsat imagery, though on extremely small scale, offers good planimetric control. Structures localising the mineralisation, dumps and workings could not be delineated but lithological differences are well seen. The aerial photos are the most easily interpretable (Fig. 4). However, due to their large number and relief displacement, planimetric control is slightly poorer. The M2S imagery, with differences in scale along X and Y directions, is very difficult to compile due to high distortion.

Optimising channel selection The interpretability of different features are tabulated in Table 1. Aerial photos offer maximum information due to the large scale and stereoscopy. The combination of two bands offer better interpretability than any single image. Among the M2S images, channels 2,6,9 and 10 individually yield better information. The overall best single image is channel 6. Among the image pairs, the 6+9 combination is the best in the interpretation of litho units, structures and mineralisation features, while the 1+2 and 2+9 pairs score over the others in the clarity of topographic features. Overall, the best M2S channel pair appears to be 6+9, with channel 1+2 coming next, followed by 9+10.

Examination of the M2S and Landsat spectral bands reveals that the M2S 6+9 combination is moreor less analogous with the Landsat 5+7 combination. The M2S spectral bands 1+2 are not represented in the Landsat. The third best M2S combination, viz. 9+10 is more or less similar to the Landsat band 7.

Conclusions An optimum choice of upto 5 of the 11 channels would give sufficient data for mineral surveys. Channels 1,2,6,9 and 10 feature among the best 4 pairwise combinaations. The band-widths found optimal here (620-660 nm, 770-860 nm, 380-440 nm, 435-485 nm, and 970-1060 nm) could be incorporated in the design of newer scanner systems for aerial and satellite operations.

Airborne MSS imagery is bound to have X-Y distortion and attempts at computer - based correction for these would be prohibitively expensive. An approximate equalisaation of scales along and across flight during formatting, by reference to base maps is advocated to furnish low distortion imagery. A mid-way stage in run-to-run compilation of interpretations could be the construction of a drainage-physiography templet (Fig. 3), corrected to mean scanner imagery scale, from aerial photos for each area, on to which the interpretation from MSS data could be transferred.

TABLE I — Interpretability Rankings on Different Image Types
(5 — Most easily identified, 0 — Not identified)

IMAGE TYPES / FEATURES	AIR PHOTO	LANDSAT MSS SINGLE BAND				LANDSAT MSS BAND PAIR						M2S AIR-BORNE MSS SINGLE CHANNEL								M2S AIR-BORNE MSS CHANNEL PAIR										
		4	5	6	7	4+5	4+6	4+7	5+6	5+7	6+7	1	2	3	4	6	9	10	11	1+2	1+6	1+10	2+3	2+9	3−6	3+9	3+10	6+9	6+10	9+10
TOPOGRAPHY & CULTURE				■	■				▲	■	▲	▲	▲			▲	▲	■		■				▲				▲		▲
RELIEF	5	1	1	2	1	1	1	1	2	2	2	2	2	1	2	2	1	1	1	4	3	2	2	2	3	3	2	3	2	2
DRAINAGE	5	2	2	2	3	1	2	2	3	3	3	4	1	1	1	3	2	2	1	3	2	2	2	3	2	3	2	4	2	3
ROADS & VILL	5	0	0	0	0	1	1	1	0	1	1	3	3	1	1	2	3	4	1	3	2	3	3	3	2	3	3	2	2	3
FIELDS	4	0	0	0	0	1	1	1	0	1	1	4	4	0	0	3	4	4	1	3	3	4	1	4	0	1	1	3	4	4
LITHO — UNITS			●	■	▲				▲	■	▲	▲	■			■	▲			●	▲							■		●
GNEISSES	5	3	3	4	2	3	3	4	5	5	5	2	3	1	2	4	3	2	3	3	4	3	2	3	3	2	3	4	3	4
QUARTZITES	5	2	2	3	3	1	2	2	3	4	3	2	3	1	2	2	2	3	2	3	3	2	2	2	3	3	2	4	2	3
LST. & SHALE	4	1	3	2	2	2	2	2	3	4	3	3	4	1	2	4	3	2	1	4	4	4	2	4	3	4	3	4	4	4
BASIC SILLS	4	1	1	2	2	1	1	1	2	3	2	3	3	2	2	3	2	2	0	4	4	3	3	3	3	4	3	4	3	3
SHALE & LST.	3	1	2	2	3	2	2	2	3	3	3	3	3	1	2	3	3	3	0	4	4	3	2	4	2	3	3	4	3	4
STRUCTURES				■	▲				●	■	▲	▲	●			■				▲	▲	▲		▲				■		
FORM LINES	4	0	1	2	1	0	0	0	1	2	2	2	1	1	1	2	2	2	0	4	3	4	3	2	2	4	3	4	3	2
FAULTS	4	0	0	0	0	1	1	1	2	3	2	4	3	2	2	3	2	2	0	3	4	4	3	4	3	4	4	3	4	3
FRACTURES	4	0	0	1	0	1	1	2	2	3	3	1	2	1	2	2	2	1	1	2	2	2	2	3	2	2	2	3	2	3
UNCONFORMITY	5	3	2	4	3	3	3	3	5	5	5	2	2	1	2	3	2	2	2	4	4	3	2	4	3	2	3	4	3	4
MINERALISATION																■	▲	▲										■	●	▲
DUMPS & WKGS	4	0	0	0	0	0	0	0	0	0	0	0	2	1	2	2	2	3	0	3	2	3	2	2	3	2	2	4	3	3
STR. CONTROL	4	0	0	0	0	0	0	0	0	1	0	0	1	2	2	2	1	1	0	2	1	2	2	1	3	2	3	3	2	2
LITH. CONTR.–ASB	3	0	0	0	0	0	0	0	0	0	0	0	2	2	2	3	3	3	0	3	2	2	2	1	3	2	2	4	3	4
LITH. CONTR.–BARYTE	2	0	0	0	0	0	0	0	0	0	0	0	0	0	0	1	1	0	0	0	0	1	0	1	0	0	0	3	2	2
OVER ALL INTERPRETABILITY			●	■	▲				●	■	▲		▲			■	●			▲								■		●

INTERPRETABILITY AMONG SIMILAR IMAGE TYPES INDICATED GROUP-WISE AS

■ — BEST ▲ — SECOND ● — THIRD

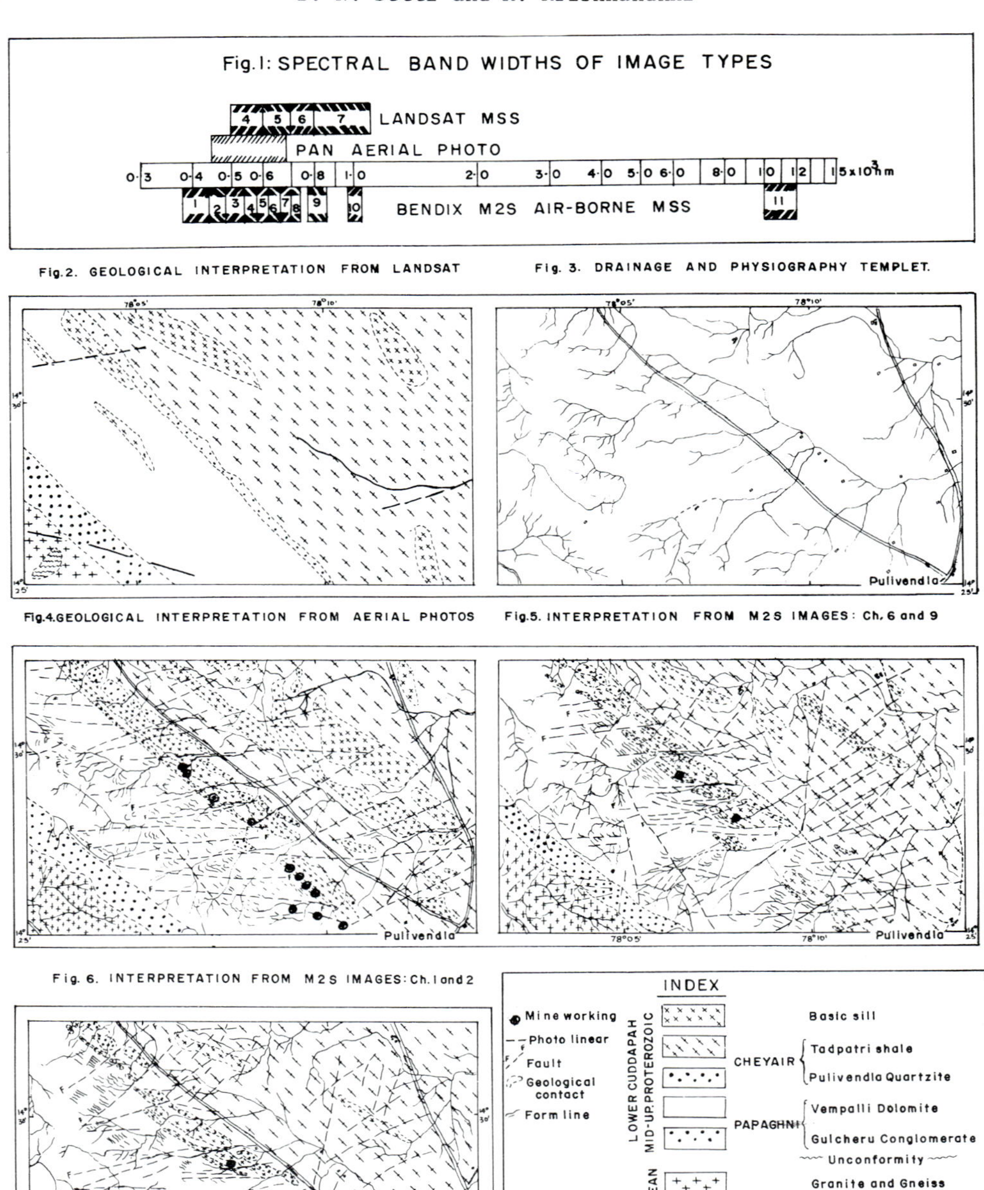
Fig.1: SPECTRAL BAND WIDTHS OF IMAGE TYPES
LANDSAT MSS
4 5 6 7
PAN AERIAL PHOTO
0·3 0·4 0·5 0·6 0·8 1·0 2·0 3·0 4·0 5·0 6·0 8·0 10 12 15x10³nm
1 2 3 4 5 6 7 8 9 10 11
BENDIX M2S AIR-BORNE MSS
Fig.2. GEOLOGICAL INTERPRETATION FROM LANDSAT
Fig. 3. DRAINAGE AND PHYSIOGRAPHY TEMPLET.
78°05'
78°10'
14° 30'
14° 25'
Pulivendla
Fig.4. GEOLOGICAL INTERPRETATION FROM AERIAL PHOTOS
Fig.5. INTERPRETATION FROM M2S IMAGES: Ch. 6 and 9
Fig. 6. INTERPRETATION FROM M2S IMAGES: Ch. 1 and 2
INDEX
Mine working
Photo linear
Fault
Geological contact
Form line
LOWER CUDDAPAH
MID-UP. PROTEROZOIC
Basic sill
CHEYAIR
Tadpatri shale
Pulivendla Quartzite
PAPAGHNI
Vempalli Dolomite
Gulcheru Conglomerate
Unconformity
ARCHAEAN
Granite and Gneiss
Basic enclaves/Dyke
SCALE
1 0 1 2 3 4 5 Km

USE OF LANDSAT DATA PRODUCTS FOR GEOLOGICAL MAPPING—A CASE HISTORY IN TAMILNADU, INDIA

V. Srinivasan, A. Sevugan Chetty, V. L. Swaminathan* and V. Tamilarasan*

State Geology Department, Madras 32, India
**Space Application Centre (ISRO), Ahmedabad 53, India*

ABSTRACT

Under a joint project of Tamilnadu Geology Branch and Indian Space Research Organization, various data products of Landsat-2 were utilized to prepare geological maps and lineament maps of Archaean metamorphic terrain in Tamilnadu. Additional information from aerial photographs and from ground checks were also incorporated for limited area taken up as test sites. The study has shown that Landsat imagery and aerial photographs can be useful as tools in the hands of an experienced geologist, even though they have certain limitations particularly in crystalline shield areas.

INTRODUCTION

With a view to measure the validity of application of remote sensing from satellites for mineral exploration work in Tamilnadu, a collaborative project was taken up with the Indian Space Research Organization (ISRO). The analysis of remotely sensed data and enhanced images from the primary data source viz., the Landsat CCT of NASA was carried out by ISRO, while analysis of aerial photographs and ground checks were carried out by Tamilnadu Geology Branch. The project area is about 60,000 Km^2 and forms the northern part of Tamilnadu. It is covered by two Landsat scenes of Path 153: Row 052 and Path 154: Row 052. The area is bound by the east longitudes 77° and 88° and by the north latitudes 10°30' and 12°30'.

OBJECTIVES

1. To delineate the different lithological units using the Landsat imagery products.

2. To classify the lineaments with reference to photogeological and ground truth data.

APPROACH (METHODOLOGY)

A number of test sites with varying amount of available ground truth and with varying nature of known mineralization were selected for study. The principal approach was to utilize the high level (Satellite), middle level (aircraft), and ground level (ground check) data in preparing geological maps for these sites in order to find out their relative ease of application, accuracy and limitations in geological mapping.

GEOLOGY OF THE PROJECT AREA

The lithological units found in the area include both sedimentary and crystalline rocks. The sedimentary rocks belong to the Tertiary and Cretaceous formations. Among the Tertiary rocks, there is a marine formation (marlstone and limestone) of Palaeocene in Pondicherry area and a terrestrial formation (ferruginous sandstone) known as Cuddalore Sandstone of Mio - Pliocene Age which occurs along the east coast overlying Cretaceous rocks. The crystalline rocks belong to the granulite suite of Archaean rocks which have been intruded by ultrabasics, anorthosite, syenite, granite and carbonatite. The granulite suite includes charnockite, mixed gneiss, amphibolite, magnetite quartzites and limestones (crystalline).

DISCUSSION

Among the Landsat imagery products used for interpretation, it is found that the False Color Composite (FCC) is better than black and white and the ratioed FCC is better than the ordinary FCC in exhibiting better tonal contrast for geological interpretation of terrain and lithological details. Among the diazo combinations tried, the three band combination with MSS 4, 5 and 7 and the two band combination with MSS 6 and 7, both in primary and complementary colours serve excellently the purpose of interpreting the geology of the area.

Geological Mapping. The geological map based on the Landsat imagery with field checks plus additional information from aerial photographs for some test sites shows the major lithological units and regional structural features including major faults. The contact 1) between the sedimentary rocks and crystalline rocks, 2) between the granites and the charnockites, 3) between the alluvium and the crystallines, 4) between the dunite with magnesite and the charnockites and 5) between the anorthosite and the granites are more or less clear in the FCC. The Cretaceous formations can be clearly distinguished from the younger formation, the Cuddalore Sandstone. The boundaries between rock units in this map, shown in Figure 1a, match with the respective boundaries in the geological map, shown in Figure 1b, prepared from conventional ground survey. This indicates that the Landsat images can be effectively used for regional geological mapping. The minor lithological units found in the conventional geological map have not been brought out in the map based on Landsat imagery. This is due to low resolution of the imagery and also due to the cover factors like soil and vegetation. Therefore, ground truth is essential and field work cannot be eliminated. However, by using these modern tools the amount of field work, the time and resources needed in a conventional geological mapping can be greatly reduced. Further, the main advantage of the Landsat imagery and a photomosaic is that they provide a synoptic view over a large extent of the earth's surface enabling an experienced geologist to interpret the geological structures which are not discernible on the ground due to the limitation of his tangential perception of a very small area on the ground. However, it is to be emphasized that experience and familiarity with the region will certainly enhance the quality of geological interpretation.

Lineaments. Many of the interpreted lineaments when they are compared with the linear features in the respective location in the photogeological maps do not indicate geological faults (see Figures 2a & 2b and Table). In an area of about 9000 Km^2, a total of 251 lineaments, 71 other linear features and 8 dykes were interpreted from Landsat imagery. After correlation, it is found that only 35 per cent of the lineaments and 7 per cent of the other linear features represent fracture and shear zones, while the rest of them represent lithological boundaries, fold axes, linear amphibolite bands, dykes, abandoned channels, linear ridges etc.

Dykes. None of the linear features interpreted as dykes is really a dyke and they indicate linear amphibolite bands, faults, linear topographic highs and linear mylonite zone (shear zones). The misinterpretations of certain linear features as structural lineaments as well as dykes may be attributed to low resolution, the cover factors like soil, vegetation and cloud and the absence of stereoscopic view in the case of Landsat imagery and the variation in the skill and experience (in the terrain of interpretation) of the interpreter.

CONCLUSION

In the joint project on remote sensing, Landsat imagery and aerial photographs were used to find out their relative usefulness and limitations in geological applications. It is found that the Landsat imagery can be used for regional lithological and structural mapping, geomorphological studies and also for groundwater study. The aerial photographs can be used for more detailed work; but field checks are essential for clearing the doubts arising during interpretation of these modern tools for identifying rock types and their boundaries. If carefully used by an experienced geologist, these modern tools can reduce the amount of field work which otherwise consume a lot of time and man power. The lineaments interpreted from Landsat imagery are to be classified with reference to photogeological data and ground truths, because many of them are misinterpreted due to certain limitations in the case of Landsat imagery and due to the interpreter's own skill and experience. After this classification, the actual lineaments representing the fracture and shear zones will be targets to the geologists for their detailed study either for mineral or for groundwater resources.

TABLE (REFER FIGURES 2 a and b)

Classification of Lineaments in Test site B1

Number in Figures	Features on Lineament Map without ground truth	Features on Photogeological Map with ground truth
1	Circular Feature	Represent a basin structure
2	Circular Feature	Folded structure
3	Circular Feature	Do not Represent any folded structure - produced by interaction of fractures and drainage divide
4 and 5	Dykes	Represent shear zones consisting of resistant rock, mylonite, forming linear ridge
6 and 7	Dykes	Topographic Feature
8	Lineament	Erosional Feature
9	Linear	Fault
10 and 11	Lineament	Fault
12	Lineament	Dyke
13	Lineament	Fault
14 to 26	Lineament	Faults
27	Lineament	Part of Major Lineament

Note Other lineaments not numbered in the lineament map have no relevance to any linear features like fold axes, faults and dykes. The East - West faults and other fold axes shown in the photogeological map are not seen in the lineament map and this may be due to scale factor, variation of the interpreter relationship (skill) and period of the Landsat image.

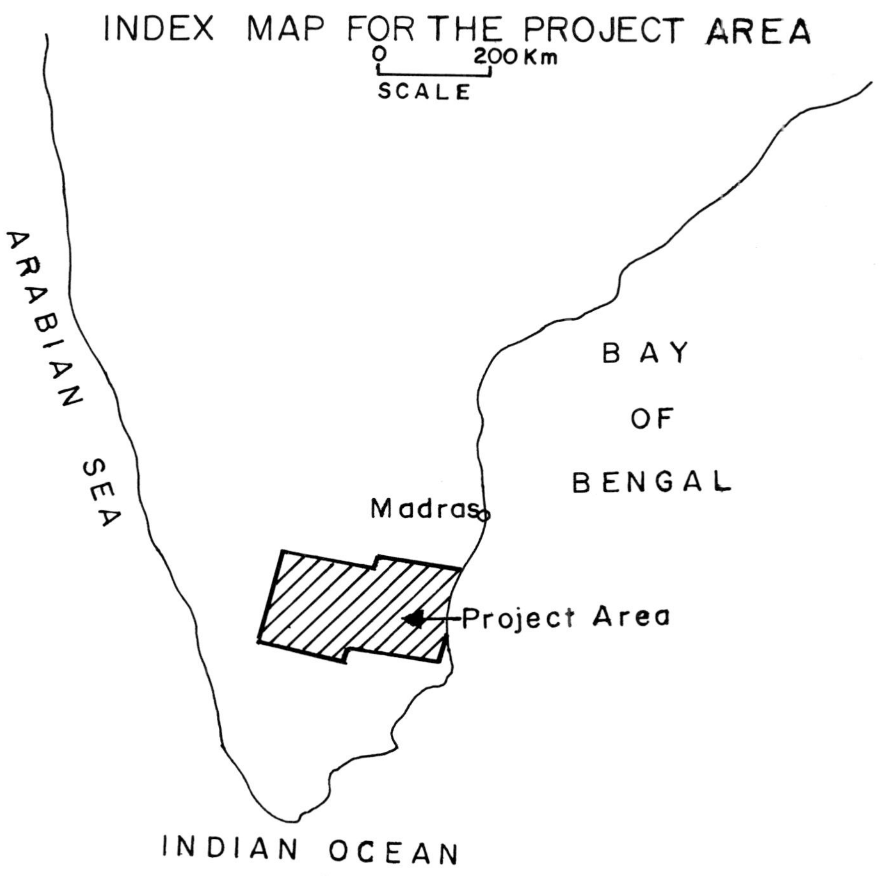
INDEX MAP FOR THE PROJECT AREA
0
200 Km
SCALE
ARABIAN SEA
BAY
OF
BENGAL
Madras
Project Area
INDIAN OCEAN

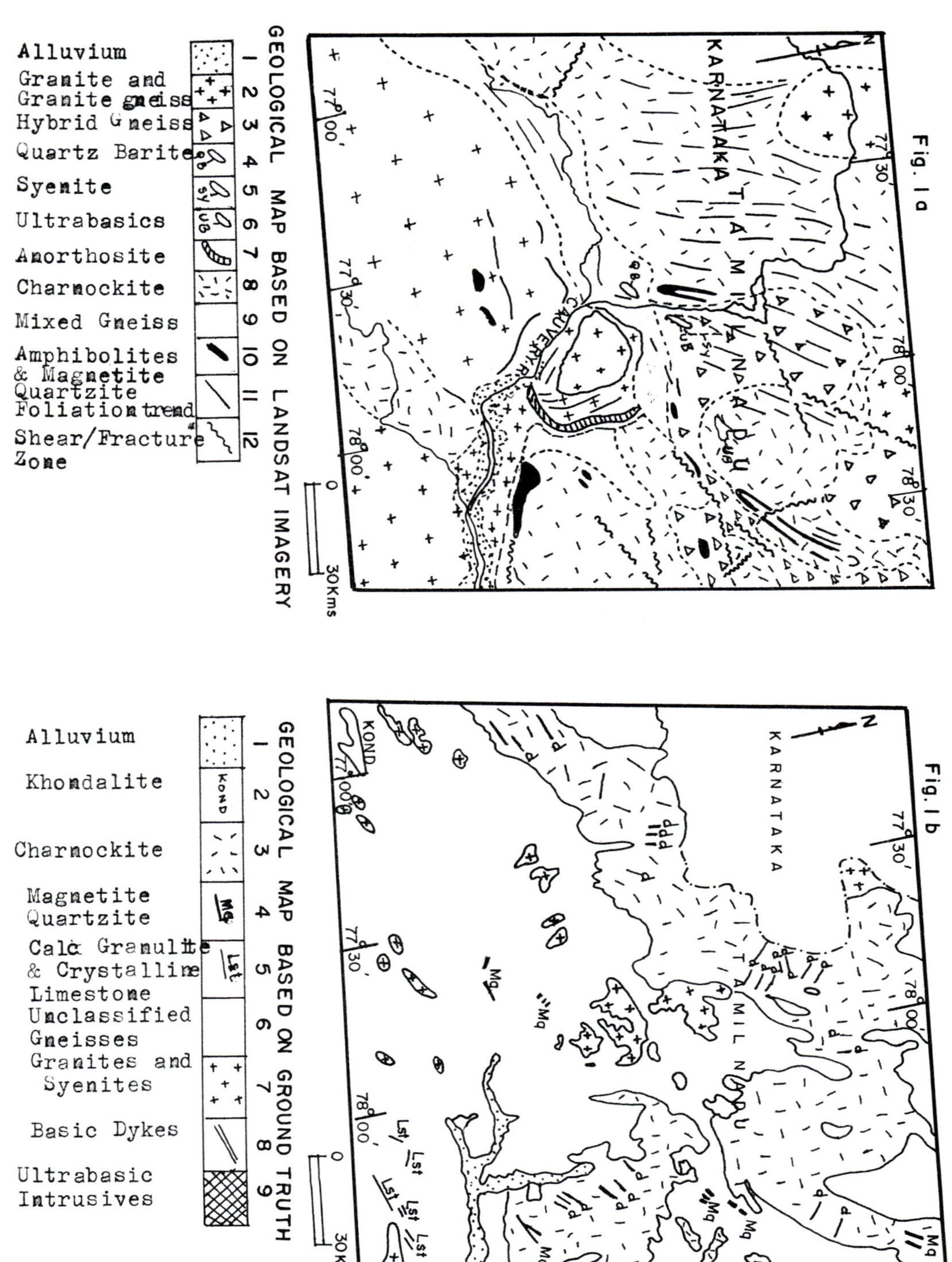
Fig. 1a
GEOLOGICAL MAP BASED ON LANDSAT IMAGERY
KARNATAKA
TAMIL NADU
CAUVERY R.
77°00'
77°30'
78°00'
78°30'
0
30 Kms
1 Alluvium
2 Granite and Granite gneiss
3 Hybrid Gneiss
4 Quartz Barite
5 Syenite
6 Ultrabasics
7 Anorthosite
8 Charnockite
9 Mixed Gneiss
10 Amphibolites & Magnetite Quartzite
11 Foliation trend
12 Shear/Fracture Zone
Fig. 1b
GEOLOGICAL MAP BASED ON GROUND TRUTH
KARNATAKA
TAMIL NADU
KOND
Mq
Lst
77°00'
77°30'
78°00'
0
30 Kms
1 Alluvium
2 Khondalite
3 Charnockite
4 Magnetite Quartzite
5 Calc Granulite & Crystalline Limestone
6 Unclassified Gneisses
7 Granites and Syenites
8 Basic Dykes
9 Ultrabasic Intrusives

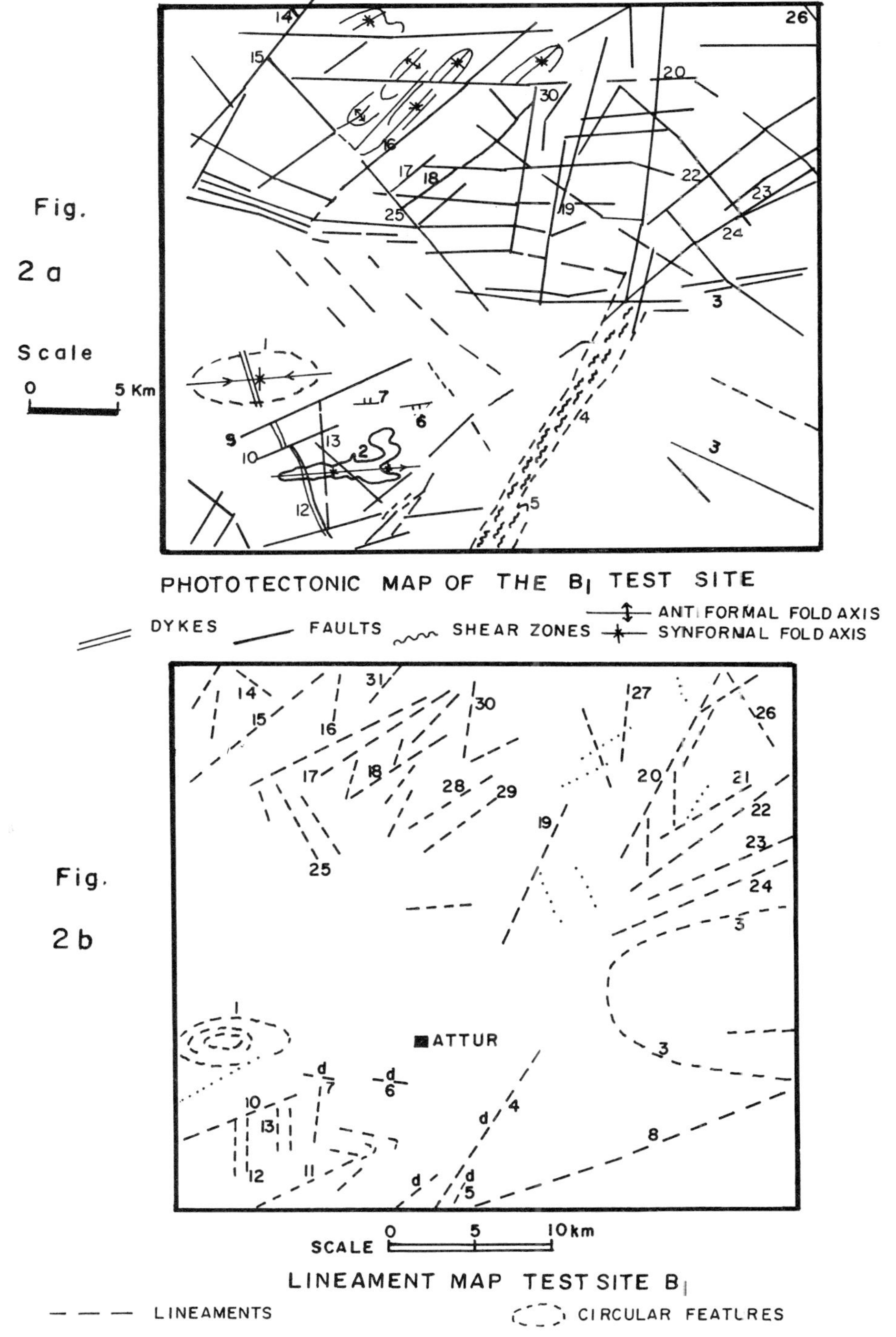
Fig. 2 a
Scale
0 5 Km
PHOTOTECTONIC MAP OF THE B1 TEST SITE
DYKES
FAULTS
SHEAR ZONES
ANTIFORMAL FOLD AXIS
SYNFORMAL FOLD AXIS
Fig. 2 b
ATTUR
SCALE 0 5 10km
LINEAMENT MAP TEST SITE B1
LINEAMENTS
CIRCULAR FEATURES
LINEAR FEATURES
1, 2, 3 Nos. IN THE TABLE FOR DESCRIPTION

LINEAMENTS OF THE NORTH CENTRAL ASIA

A. V. Ilyin

Institute of Lithosphere, Academy of Sciences, Moscow, U.S.S.R.

The present report deals with lineaments observed on space image in the Southern Siberia and Northern Mongolia. The general tectonic pattern of the region remainds to some extent that of the Northern India and adjacent territories. The area from Siberian to Indian platform seems to belong to the common field of the tectonic strain. It is caused by the Indian platform moving northward (Molnar, Tapponier, 1975). The whole area is generally alike in relation to neotectonic movements, which were sharply expressed from the Baikal rift in the north to Hymalaya Mountains in the south. More than that, in a hystoric time some events such as demolishing earthquakes happened nearly simultaneously throughout the area (e.g. in 1861, 1905, 1957).

The region under consideration has not yet been studied in detail. Some important findings were made here during current decade. Two of them are outstanding: (1) Khubsugul phosphate basin of the Latest Precambrian-Cambrian age and (2) large Erdenet porphyry-copper deposits connected with Upper Paleozoic-Lower Mesozoic volcano-plutonic belt. The location of the deposits was found to be controlled by lineaments, seen on the space image.

On the whole, space image application for the large region studied seems to be most helpful. It gives a rare opportunity to examine the known facts, to interpret mineral deposits position and to get new ideas about possible location of deposits to be discovered.

The map of lineament was compiled as a result of viewing small-scale space image taken by "Meteor" scanner system (Fig. 1). A complicated set of lineaments observed on the image generally correspond to four directions - latitudinal, longitudinal, north-west and north-east. The most clearly observed are lineaments expressed with fissures of newly demolishing earthquakes. Faults which have been found by geological mapping are also clearly seen in most cases. Generally lineaments seen on the image correspond to faults.

The region in question embraces two main tectonic units - Precambrian Siberian platform or old continent and Lower Phanerozoic orogenic belt or oceanic terrains. In most cases the bordering zone between two units is seen on the image. It is rather often that the zone is occupied with orogenic granite masses generated due to an interac-

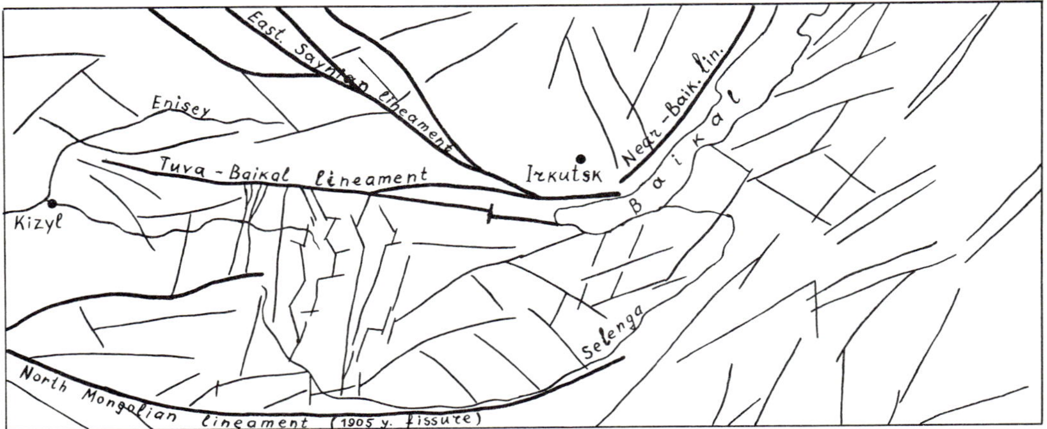

Fig.1. Work map of lineaments seen on Meteor imagery - Southern Siberia, Northern Mongolia.

tion of two old plates. Granite seems to camouflage the old continental edge, preventing it to be seen on the image, while it had been marked out with geological mapping.

On the other hand, space image reveals some new lineaments or faults which had not been known before. These are mostly transverse fault cutting the buried Precambrian basement within the old continental margin. It is especially notable that such fault were found under Latest Precambrian sediments, covering the basement on the continental margin. As for the faults parallel to folding these seem to be longer on space image than on geological maps and united in a new fault pattern. All the lineaments could be divided into three systems.

The 1st system consists of two largest zones - Eastern Sayanian and Near-Baikalian, running NW and NE correspondingly and bordering the Archean basement of the platform from the south. Both zones are rather wide and could be distinguished even on very small scaled images.

The 2nd system embrace two latitudinal lineaments some 300 km apart (Fig. 2). The southern lineament is readily observed due to a fissur originated by an earthquake in 1905. It runs as a straight line for almost 400 km. A narrow strike-slip-rift often developes along the line. Both sides of the rift are marked with small lakes, probably generated after thick permafrost cover had been fractured by the earthquake (Ilyin, 1978). This newest fissure turns out to be strictly dependent on the oldest structural pattern, since it runs along the Latest Precambrian continental edge, fixed on the surface as an out er border of carbonate facies of the Latest Precambrian shelf. The same line happens to be an axis of latitudinal ensialic volcano-plutonic belt, Upper Paleozoic-Lower Mesozoic in age. In one of the cases, where the main axial lineament crosses the smaller one the large Erdenet porphyry-copper deposit happens to encounter.

The northern lineament was not known before. Instead the mapping revealed several faults isolated each from the other by Cenozoic basalt flows, by glacial accumulations and Lower Paleozoic sediments Viewing the space image one can readily join these faults into a lon lineament running west of Baikal.

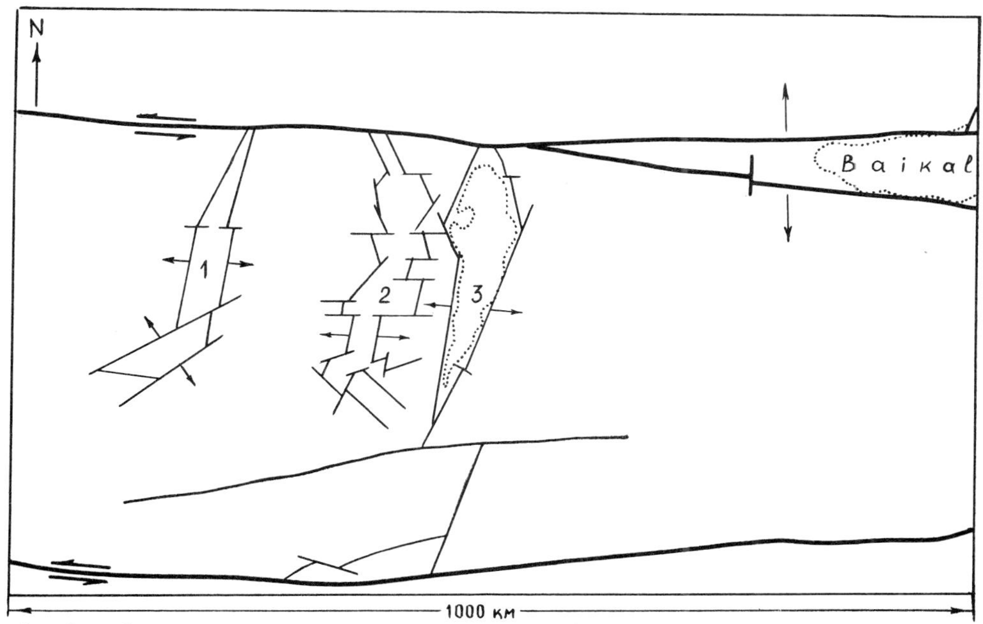

Fig.2. The western closure of the Baikal rift system (Rifts: 1 - Busein, 2 - Darkhat, 3 - Khubsugul).

Fig.3 (below). The Darkhat rift. Black spots - alkaline intrusions with figures designating K/Ar age in m.y.

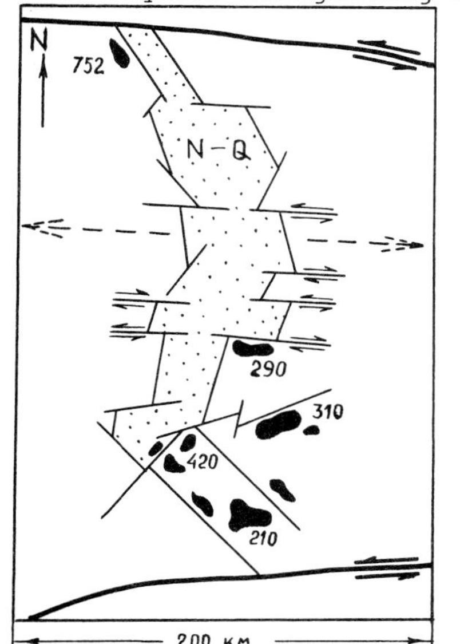

The 3rd lineament system consists of three sets of longitudinal faults, located between latitudinal lineaments of the 2nd system. Each of the set comprises borders of Cenozoic rifts filled with thick clastic piles.

The easternmost rift or Khubsugul lake is about 150 km long, and 10 to 40 km broad. Being a sharply expressed feature of the present day relief the Khubsugul lineament zone is also the oldest weak zone, most probably coincided with the inner border of the Latest Precambrian-Cambrian shelf. At that time the area west of the zone was subsided, and that to the east was raised. In Cenozoic the movement along the same zone takes an opposite course. In that capacity the Khubsugul zone remainds one of the Las Vegas Hinge line in the Western United States (Stewart, Pool, 1974).

The Darkhat rift was also occupied by a large lake. The western side of the rift was cut through to the bottom by the Enisey river not long ago and the lake was drained. Therefore on space images and on land an observer is able to see how the rift is really constructed. It is a typical block structure. A number of blocks can be guessed along the axis of the rift, since its bottom consists of alternating hills and flat low areas. These are separated with linear features. Some of the lineaments are seen in places, where thick Cenozoic accumulation is known to exist.

In a plan view the rift has a rectangular outline (Fig. 3). It was generated as a result of crustal extension. But the extension strain was performed by a motion along the transverse shear zones, which has long been existed in the basement. With the extension strain, blocks located on both sides of the rift were pulled apart in opposite direction in a manner, proposed by Abdel Gawad and Tubbesing (1976).

Transverse latitudinal lineaments corresponding to shear zones continue to horst massives adjacent to rifts. They are accompanied there by Paleozoic alkaline intrusions with rare metal mineralization. Alkaline intrusions vary in absolute age from Lower Paleozoic to Middle Mesozoic. It shows unmistakably throughout time a persistence of tectonic tension during most of the Phanerozoic. As for lineaments bordering rifts they seem to be old in origin too and determine the tectonic and sedimentation pattern from Latest Precambrian. Phosphate facies of the Latest Precambrian-Cambrian are controlled by sea bottom rises and inner shelf border, which in their turn are strictly dependent on lineaments.

The lineaments happen to be sites of high-discharged thermal springs of mineralized water used for medical treatment. They also serve as paths for sweet water circulation and a lot of big ascending springs are confined to lineaments.

To interpret lineaments of the 2nd and the 3rd group one should bear in mind the Baikal rift system, laying just to the east. Two longitudinal lineaments can be viewed upon as large strike-slip faults, connecting the Baikal rift system with smaller rifts described above. Thus a united rift system including Baikal and smaller rifts, appears to be developing in Southern Siberia and Northern Mongolia. Again extensional strain is performing in a much broader scale through transverse shear zone represented with lineaments of the 2nd system.

To conclude, space image in the North Central Asia can be used for better understanding of the old structural pattern, for mineral deposits exploration, for water (thermal water) supply. The new mode of generation of Cenozoic rifts can be proposed viewing the image. Transverse faults accompanied with alkaline intrusions can also be located on the image. The space image helps to delineate the distribution of phosphate facies and to direct exploration for old phosphorite. The location of new deposits can be predicted using those already known on one hand and lineament pattern seen from the space on the other.

REFERENCES

1. P.Molnar, P. Tapponier, Science, 189, 419-426, (1975).

2. A.Ilyin, Earth and Planet Sci. Let., 41, 107-109, (1978).

3. J.H.Stewart and F.G.Pool, in: Tectonics and sedimentation, Soc. Econ. Pal. Min. Sp. Pub., N 22, 1974, 28-57.

4. M.Abdel-Gawad and Linda Tubbesing, in: Proceedings of the First International Conference on The New Basement Tectonics, Salt Lake City, 1976, 61-81.

SOME RESULTS OF REMOTE SENSING IN YUGOSLAVIA

B. Koščec

Industroprojekt, Department of Complex Geological Exploration, Zagreb, Yugoslavia

ABSTRACT

A short history of application of remote sensing technology and methods in Yugoslavia is given in the paper, with special emphasize to geological and affiliate purposes. Some case histories of successfull application of LANDSAT images, standard issues and computer enhanced, in exploration of different mineral resources in different regions of Yugoslavia are briefly discussed. Thermal infrared scanning has been carried out in order to locate possible mineralized zones, to check some geothermal anomalies, to solve certain coastal hydrogeological and water pollution problems on Adriatic shore as well as to perform the energy-loss survey for industrial and urban purposes.

INTRODUCTION

Airborne photography in Yugoslavia dates back to 1923. However, until 1945 it was used mainly for military purposes. Ever since black & white aerophotography has been used extensively, first in photogrammetry and cartography and then (about 1950) in forestry, geology, agriculture and other fields of activity. Geological applications, basically for mapping purposes, were started at the University of Belgrade (M. Dimitrijević, B. Stepanović) and, in 1962, in Zagreb (Industroprojekt Co.), to become a must for all geological institutions in Yugoslavia involved in the production of the Basic Geological Map (1:100,000).

Other airborne photo techniques, such as color, black & white, infrared, and color infrared aerophotography were introduced in Yugoslavia in 1970 by the geologists of Industroprojekt Co. in Zagreb. Since then IR aerophotos have continuously and successfully been used in tackling a variety of engineering geology problems.

Since 1973 Yugoslav geologists have been using, on a continuous basis, the satellite images of the LANDSAT family (started with ERTS-1) in regional studies of tectonic--structurological relations as well as in studies related to exploration for water, oil and other mineral resources. A photomosaic of all Yugoslavia, based on ERTS-1 images, was produced in 1974/75 (Fig 1). Advanced techniques, such as computer enhancing of LANDSAT images meant to improve analyses of morphostructural features, were first applied in 1975 through the U.S.G.S. Eros Data Center in Sioux Falls. Today this is a common technique in preparing satellite images for photointerpretation. Computer processing is still run at Remote Sensing Center in Oberpfafenhofen and Munich (West Germany), but every effort is made to establish computer processing facilities in Yugoslavia (probably in Zagreb) as soon as possible.

SKYLAB images reached Yugoslavia in 1974. Some of them were of very good quality and tectonic analyses were performed for the bauxite-bearing coastal area (Industroprojekt, Zagreb) and the north-western part of Yugoslavia-Slovenija (Mining & Geological Faculty, Belgrade).

The airborne multispectral and thermal infrared scanner was introduced in Yugoslavia in 1974, and several multipurpose (mainly geological) test flights were made. The thermal IR part of the system (by Daedalus Inc., U.S.A.) became operational in September 1976. Following the experimental phase and initial problems, some very acceptable results were obtained in the field of mineral resources, geothermal exploration, coastal hydrogeological studies, water pollution and heat-loss surveys.

APPLICATION AND RESULTS

A very good SKYLAB image of the coastal area (Obrovac-Šibenik) made possible the interpretation of lineaments, which is very important for an adequate determination of the structural framework of this bauxite-bearing region (Figure 2). The prevailing fault system is of the "Dinaric" type (NW-SE) but nearly all of these faults are intersected by younger transversal or diagonal systems a few kilometers long (Oluić, 1978). These younger faults, clearly detectable on the SKYLAB image, displace the longitudinal structures and disrupt the continuity of lithological units. Two basic generations of faults have been recorded: the older, in Cretaceous sediments which dip under Paleocene deposits, and the younger, affecting both Cretaceous and Paleocene sediments. Bauxite deposits are bound to some of these faults. The photointerpretation was made by INDUSTROPROJEKT - Zagreb. Another good SKYLAB image of the north-western part of Yugoslavia (Slovenia) has been interpreted by M. Dimitrijević from Belgrade University. Some very important lineaments of seismic importance have been determined, as well as tectonic-structural conditions favouring exploration for different mineral resources. By using the LANDSAT image of the same territory (Slovenia and adjacent areas) INDUSTROPROJEKT geologists (M.Hanich et al.) not only confirmed the results mentioned above but extended the lineaments to a broader area and recorded some ring-shaped structures of great importance for further investigation of ore deposits.

Photointerpretation of the LANDSAT 1 image of the area west of the city of Zagreb (border area between Croatia and Slovenia) confirmed the already known faults and disclosed some new hitherto unknown lineaments (Oluić, 1975). Three basic directions of the fault systems were indicated in the area, the NE-SW direction dominating. The zones where these systems meet are most intensively disturbed which explains the low seismic stability of the area under consideration (Figure 3).

The occurrence of thermal springs in this area is connected with some lineaments of NE-SW direction. The probably most popular recreation center near Zagreb, Cateške Toplice thermal springs, was flown by the thermal infrared scanner system in early 1974 (Figure 4). Besides the higher thermal intensity of the swimming pools with warm water (36 ^{o}C), some hitherto unknown thermal anomalies were recorded in the lower part of the discharge channel, in the nearby fish-pond and in the river Sava. When these thermal images were correlated with the results obtained by photointerpretation of airborne photos and LANDSAT images, it was found that these thermal anomalies fully match the strong fault line of the same direction.

Computer enhanced LANDSAT images, as already mentioned, were used for the first time in 1975 in order to increase contrast and thereby express better the differences in the imaged area (Hanich, B.Koščec, Denih 1978). The chosen area covered the exploration zone of the Bor and Majdanpek copper mines in eastern Yugoslavia. Digital processing of the LANDSAT image of the area has been performed at the U.S. Geological Survey, EROS Data Center in the U.S.A. in cooperation with Yugoslav geologists who have subsequently carried out the photogeological analysis and interpretation of processed images.

Besides the three prevailing fault systems in this area, the photogeological analysis of the satellite image recorded a number of ring structures whose contours can be noticed despite their being covered by younger structures or thick Tertiary sediments. Some of these structures feature a more or less prominent rim of the ring with typical subsidence of the central part inside the ring, wherease others have a clearly raised central part (Figs. 5 and 6). The reason why these structures can be seen on the surface is primarily to be found in the fact that a satellite image makes it possible to relate phenomena which cannot be detected by field prospecting or by an analysis of airborne photography, because of the wide area in which they occur. Since the occurrences of effusion, mineralization, and thermal water were found to be related to the ring-shaped structures - in most cases to their peripheral parts - identified by photogeological analysis, the need has arisen to subject areas where such structures occur to a more detailed analysis. This fact, as the most important guideline for future exploration programming in this area, represents a very important result of the work mentioned above.

A year earlier, in April 1974, one part of this area, near the Bor copper mine, was flown by multispectral and thermal infrared scanning systems in order to test the system in geologically a more or less known area (J.Koščec & B.Koščec, 1975). Some 42 square kilometers of the so-called Timok Magmatic Complex, consisting mainly of igneous rocks with hydrothermally and metasomatically altered portions, were covered (Fig. 7). Higher thermal effect were yielded by altered volcanic rocks (silicified, pyritized and chloritized), Upper Cretaceous clastic rocks and andesites. Altered rocks, however, which are the bearers of copper mineralization in the studied area, displayed the highest thermal effects, some of which pointed to ore-bearing zones located during previous geological, geophysical and geochemical exploration. On the other hand, three of such locations, recorded by this flight as highly promising in terms of existing thermal effects (no other indications were known by that time) turned out to be copper bearing as determined by geological and geophysical investigations carried out in 1975 and 1976.

LANDSAT 1 and 2 images have been used by the geologists in the Republic of Bosnia and Hercegovina in order to clarify the tectonic and structurological relations, as an important sequence in their work on the metallogenetic map of the territory. Work on a study involving mineral resources, earthquake manifestations and geothermal energy of the said area is currently under way, and one of the basic steps in this study has been photointerpretation of aerophotos and LANDSAT images. Neotectonic studies related to the exploration of geothermal resources and mineral deposits in Slovenia, as well as the work on the Yugoslav Basic Geologic Map scale 1:100,000 in Serbia, included photogeological analyses of aerophotos and LANDSAT images as a significant part of the work.

After the photointerpretation of LANDSAT 1 and 2 images covering the territory of the Republic of Macedonia (Dimitrijević & Marković, 1976), which provided new knowledge on the geotectonic framework and structural-geological relationships of the area, satellite images are still currently used in local and regional seismological projects.

Petroleum exploration in the southern part of Pannonian basin was supported by a photogeological analysis of computer enhanced LANDSAT images (1977/78), which yielded significant results in the solution of tectonically related problems of the area.

In Macedonia, the southeastern part of Yugoslavia, thermal infrared scanning was carried out during the winter of 1976 and the obtained results interpreted in early 1977. The goal was to explore mineralized zones where other geological and geochemical prospection work was carried out simultaneously, but separately (J.Koščec, 1978). In the geological sense the area is built up of sedimentary, igneous and metamorphic rocks of Paleozoic, Mesozoic, Tertiary and Quaternary age. Because of multiphasic magmatism and metamorphism a series of minerals and ore occurrences (such as pyrite, chalcopyrite, magnetite, titanomagnetite, antimony, arsenic, galenite and sphalerite) are present.

Analysis and interpretation of the results provided some very useful information, such as:

- the tectonic and structural elements predominating in the area are ruptures and ring-structures (Fig. 8);
- because of lithological composition, geological boundaries in sedimentary and metamorphic rocks were rather easily determined (limestones, sandstones, granites, metadiabase etc.);
- hydrothermally altered (silicified, pyritized and limonitized) rocks are characterized by higher thermal effects (Fig. 9);
- in terms of their location and shape, most thermal anomalies correspond to geochemical anomalies determined by field geochemical exploration.

The thermal infrared scanner was also used in 1977 in the Northern Adriatic coastal area (Quarner Bay) to detect the present situation in terms of water pollution and submarine fresh water discharge. Since this is an industrial (port of Rijeka) and tourist area it suffers both from pollution problems, which tend to grow with the development of tourism and the petrochemical industry, and lack of fresh water on the karst coast and on the islands (Knapp & B.Koščec, 1979).

The results of this study confirmed the location of huge fresh water discharges into

the sea and of the main existing pollutants, and provided an approximate estimate of pollution intensity (Fig. 10).
Similar studies have recently been extended to other parts of coastal area.

Encouraging results were obtained in 1978 in detecting heat-loss by thermal infrared scanning of steel mills and some new urban developments in the Republic of Slovenia. In 1979 this work will be extended to some other urban areas.

In order to bring together researchers and experts of all disciplines dealing with remote sensing, the Yugoslav Academy of Sciences and Arts in Zagreb has recently established the Council for Remote Sensing and Photointerpretation which, among other tasks, will coordinate remote sensing activities and cooperate with related associations and individuals abroad.

References

1. M. Dimitrijević and M. Marković, Symposium on Remote Sensing, Belgrade, 1976.

2. M. Hanich, B. Koščec and M. Denih, Space Research, v. XIX (1978)

3. M. Knapp and B. Koščec, Proceedings of II Conference on Protection of Adriatic Sea, Book II, 97, Hvar, 1979.

4. J. Koščec, B. Koščec et al., Proceedings of Tenth International Symposium on Remote Sensing of Environment, 911, Ann Arbor, Michigan, 1975.

5. J. Koščec, M. Knapp et al., 12th International Symposium on Remote Sensing of Environment, Prog. Abs., 257, Manila, 1978.

6. J. Koščec, R. Radaković and M. Denih, II Yugoslav Consultation on Lead - Tin Mineralization, Štip (Macedonia), 1978.

7. M. Oluić, Geološki Vjesnik, 28, 87, Zagreb, (1975).

8. M. Oluić, Geološki Vjesnik, 30/2, 731, Zagreb, (1978).

Figure 1: The LANDSAT 1 photo-mosaic of Yugoslavia

Figure 2: SKYLAB - 3 image of the coastal area (bauxite-bearing region)

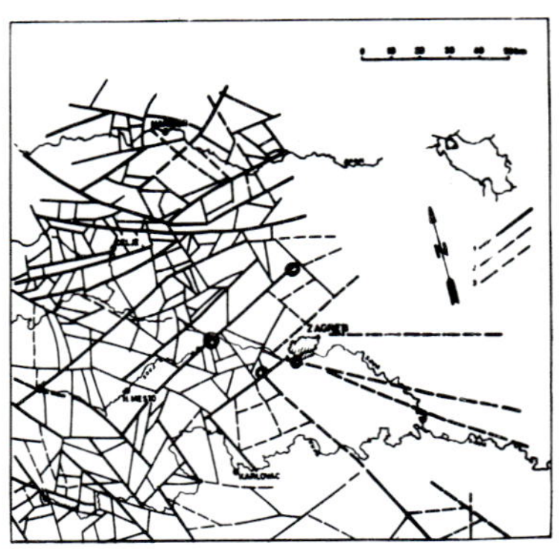

Figure 3: Tectonical analysis based on LANDSAT 1 image No. E - 1198-09221 (black dots = thermal springs)

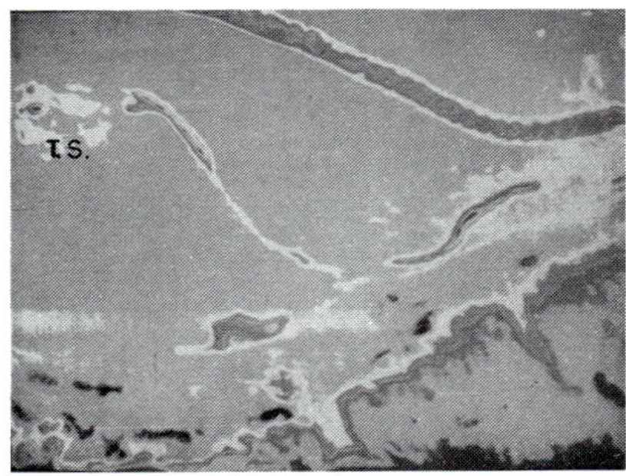

Figure 4: Thermal infrared image of Čateške Toplice (T.S. - Thermal springs)

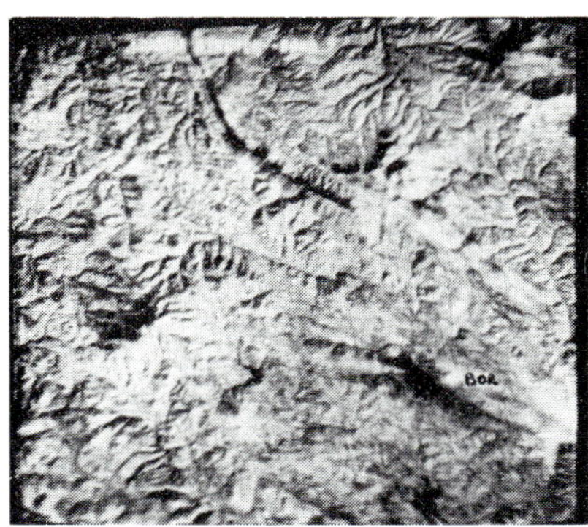

Figure 5: Computer enhanced LANDSAT 1 image No. E-1463-08511 (Bor copper mine area)

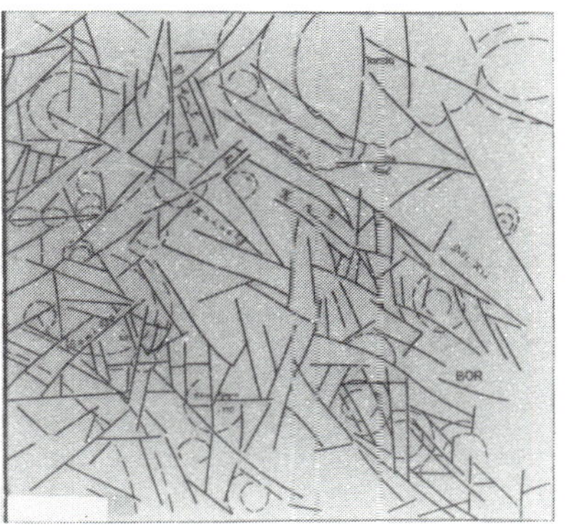

Figure 6: Tectonic - structural study of the area shown on figure 5

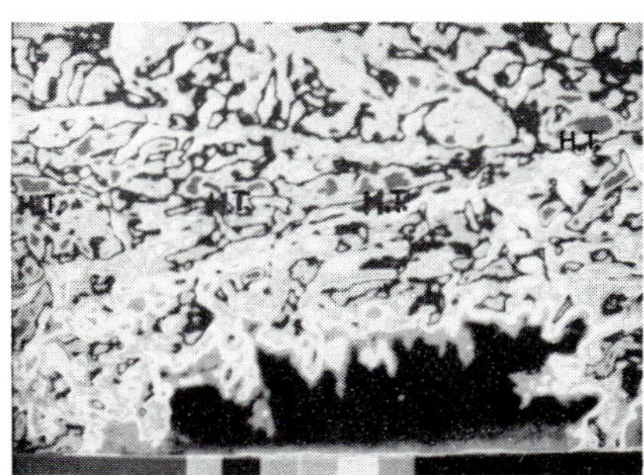

Figure 7: Thermal effects recorded by thermal IR scanner in copper exploration area north of Bor mine (H.T. - higher thermal effects)

Figure 8: A ring-shaped structure registered by thermal IR scanner in Macedonia

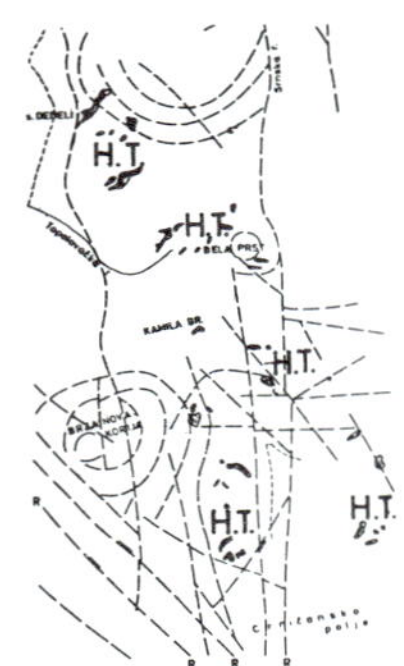

Figure 9: High thermal effects (H.T.) in an exploration area in Macedonia (TIR scanner)

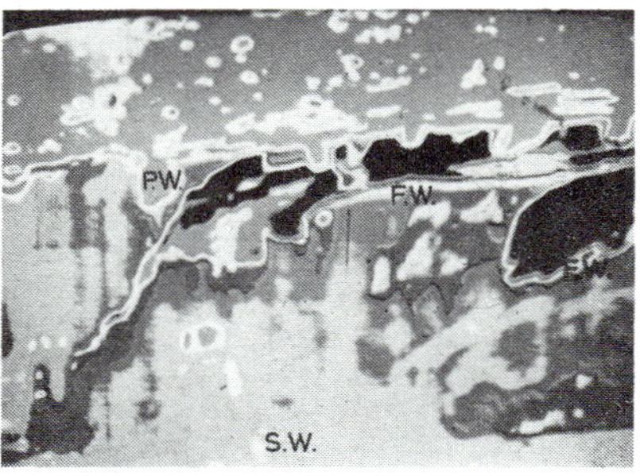

Figure 10: Fresh water and polluted water discharge in the Adriatic Sea (port of Rijeka)
S.W. - sea water
F.W. - fresh water
P.W. - polluted water

APPLICATION OF LANDSAT-2 DATA FOR OBTAINING LAND INFORMATION

R. Haydn, F. Jaskolla and J. Bodechtel

Zentralstelle für Geophotogrammetrie und Fernerkundung (ZGF), München, Federal Republic of Germany

1) INTRODUCTION

Technical cooperation is not very effective, if technology is being provided exclusively on the basis of scientific hardware and equipment.

For beneficial cooperation, we should help to improve in the long run the quality of life, or at least to create acceptable fundamentals for living, on the basis of the given natural resources. The expression "natural resources", used in relation to the energy crisis which we have been facing for several years, has to a great extent been associated only with the so-called non-renewable resources. A profound scientific understanding of our environment with respect to the renewable resources is also an indispensable requirement. The lack of thematic earth scientific information and the lack of scientific experience in information acquisition and its beneficial transformation into useful applications is a fact which, unfortunately, applies especially in the developing areas of our planet.

Since the availability of the LANDSAT-System - the first experimental space platform exclusively designed for earth observation purposes - we have for the first time a tool by which we are able to collect thematic earth scientific information on a global basis and also in those areas which are difficult to access. The application of LANDSAT-data for development aid purposes can be seen under operational aspects if certain requirements in data handling and utilization are taken into consideration.

The purpose of this paper is to demonstrate a strategy in utilizing space borne data on a routine basis for the production of thematic earth scientific information.

2) OVERALL CONCEPT FOR THE OPERATIONAL APPLICATION OF LANDSAT DATA

It is a matter of fact that LANDSAT data cannot fully replace any conventional methods of deriving thematic information. But it can support to a large extent mapping activities on a small scale basis. The overall concept which has been used for mapping purposes is utilizing all available informations consisting of LANDSAT CCT's, thematic maps (if existing) and additional literature. On the basis of these input data information preprocessing may be carried out in order to prepare a preliminary interpretation on satellite images.
The most critical point is the methodology for combining image processing and subsequent interpretation in an optimum way. A possible approach is demonstrated by fig. 1.

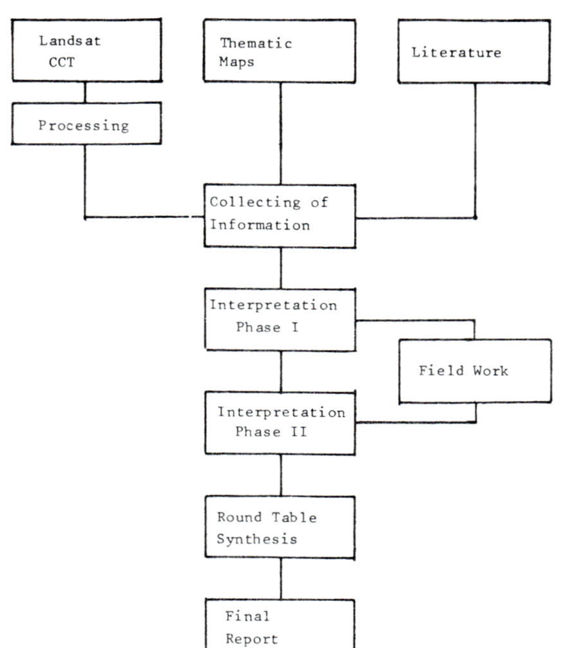

Fig. 1:

Overall concept for the operational application of LANDSAT data

After collecting all above information for the preliminary interpretation (Phase I) various digital preprocessing and enhancement techniques as described in figure 2 have been applied in order to produce optimized images for photo-interpretation activities.

On the basis of the preliminary interpreted thematic maps a verification of the results was carried out by a field trip which could be restricted to a minimum of time. The final results as basis information for further cartographic activities can be produced by a synthesis of the newly gained information including ground truth data gathered until then. This approach of incorporating various information sources and the full range of digital image processing with conventional methods of photo-interpretation has operational aspects and enables us to carry out small scale inventories over large areas within a reasonable period of time.

Based on our experiences we can state that the utilization of automatic classifiers cannot be regarded as an optimum approach even under quasi operational aspects. This is to a far extent based on the fact that at least the available spectral information is not sufficient to accurately describe the manifold natural environment.
Best results could be obtained by applying photo-interpretation techniques on computer enhanced image products such as linear (Principal Components) and non-linear (Ratio) transformations including various filtering processes.

The definition of best suited parameters for the corresponding processes has to be carried out under the control of an experienced interpreter. This can easily be achieved by processing test areas under various aspects before transforming the entire data set.

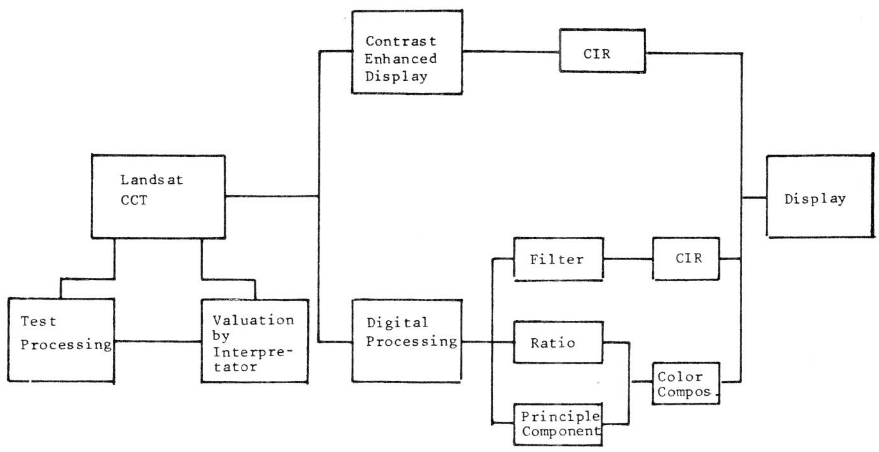

Fig. 2: Scheme of digital image preprocessing

3) APPLICATION OF THE OVERALL CONCEPT

The benefit of using satellite data for obtaining land-use information regarding above overall-concept can be demonstrated by an investigation which was carried out in the Islamic Republic of Mauretania.
The test-site, the so-called Targant Region is located at the northern boundary of the Sahel Area between 13° and 15° West and 17° and 19° North. Two different types of landscape characterize the Targant: In the eastern Tidjikja Plateau a peneplain dipping slightly to NE forms the surface. It is crossed by rivers draining to the north. The western part consists of intensively jointed and folded rocks. It is drained to the west by two major systems which end in the depression of Gabou.

The interpretation i.e. the differentiation of relevant surface phenomena was carried out in the scale of 1 : 2oo.ooo. The investigations in the test site of about 42.ooo km^2 were finished within 4 months (1 manyear) including a field trip of 4 weeks.

In order to use the whole range of information of LANDSAT data, 4 different maps of surface phenomena were produced:

- Geological Map

Similar to the conventional aerial photointerpretation different photogeological units based on greytone- and texture variations were selected.
All together 14 different units could be selected and mapped by comparing with the available geological map. Within these units further differentiations are possible. To differentiate the various geological phenomena, several digital processing products were used.
Fig. 3 shows a Ratio of the channels 4 and 5, where different kinds of dunes and flying sands are easily interpreted.

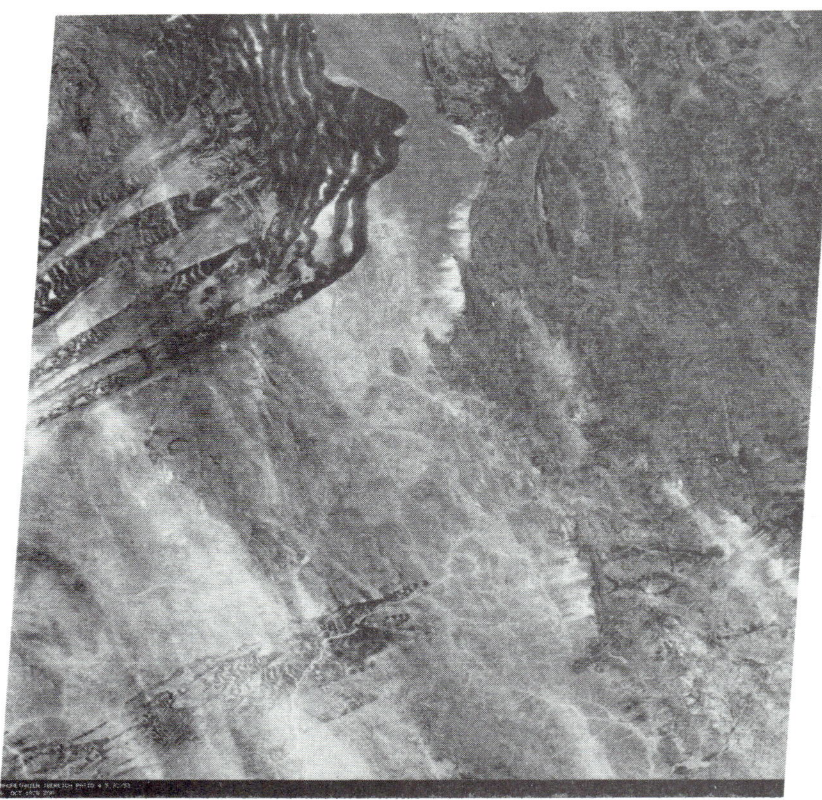

Fig. 3: LANDSAT-2 image of the western Targant Region (Moudjéria) Ratio of the channels 4 and 5

- Vegetation Map

Beside the possibilities of the differentiation of vegetation by use of false-color-composites good results could be obtained interpreting by use of a special components transformation.
In a first step the vegetation was classified in density categories. Regarding figure 4, which shows the transformation of the same area as in figure 3, dark and small structures can be separated adequate to the dense and cultivated vegetation in the oasis. Further density-changes were interpreted visually.

- Soil Map

On the basis of various processed images it was possible in a first step to delineate homogeneous areas with respect to the spectral behaviour. In order to obtain indications on soils and soil-associations information on geology, vegetation, drainage pattern and field investigations had to be incorporated.

- Map of Potential Land Use

After the production of all above information, finally a map of potential land use was made available. The cartographic description expresses those areas, which are worthwhile for further developing activities (Feasibility Study).

This map has to be regarded as the synthesis of all available and generated data and serves for further planning activities.

Fig. 4: LANDSAT-2 image of the western Targant Region (Moudjéria) Principle Component Transformation

The work was carried out by a consortium of Zentralstelle für Geophotogrammetrie und Fernerkundung - ZGF - and Institut für Planungsdaten - ifp - for the German Agency for Technical Cooperation.

MEASUREMENT RESULTS AND CONCLUSIONS ON THE SPECTRAL REFLECTIVE COEFFICIENTS OF VOLCANITES, GRANITOIDES AND GNEISSES

H. B. Spiridonov, A. H. Krumov, N. K. Katzkov*
and S. R. Yovchev*

Central Laboratory for Space Research
**Applied Research Laboratory, Sofia, Bulgaria*

ABSTRACT

The simple interpretation of different genetic classes of rock formations when using air- and space-techniques is still one of the unsolved problems of remote sensing in the field of geology. The visual and instrumental interpretation of multizonal photographs and photoelectron pictures, regardless of success attained, is far from the final solution to the problem. A pattern of reasonable solution is the measurement of the spectral reflective coefficient (SRCO) of rock formations in their various states, accumulation of statistical data and compiling data catalogues to be used further in computing the space information obtained. This paper shows results obtained from ground-based measurements of SRCO of representative magmatic and metamorphic rock samples, taken from various parts of the country. Based on this, some conclusions were made, to provide a reasonable background for future studies in this field.

DEFINITION OF THE PROBLEM

Remote sensing of the Earth by air- and space-borne highly sensitive instruments provides images of land and water surfaces in several ranges (channels) of the visible and near infrared portion of the electromagnetic spectrum. These images are of general application and the different formations are outlined within the respective channels. Based on this information, geological and geomorphological studies specify the location, shape and sizes of various geological bodies; and maps of lineament and circular structures may be compiled. All this is an initial stage in applying remote sensing to geology. In the future there have to be developed methods of identifying geological objects, based on their spectral reflective characteristics (SRC). A necessary condition here is to provide for systematic and multiple laboratory and ground (field) studies on the main classes of geological objects at their different states and to compile a catalogue of SRC. Some work has been done in our country in cataloging the SRC of the major genetic rock classes: magmatic, metamorphic and sedimentary.

INSTRUMENTS AND MEASUREMENT TECHNIQUES

Our Laboratory has developed a portable spectrometer to measure the SRC of various earth formations and objects in vitro and in situ. The instrument, known as ISOCH-020 is of twenty channels with channel width of 10-12 nm and operates within the range of 400-800 nm at aperture angle of view - 13°. The natural sunlight is used in field work and in laboratory conditions - an artificial light source. A specially designed screen with photometric characteristics close to the one of ideal reflective white surface is used as reference object.

The rock samples to be studied have the shape of irregular polyhedrons with natural roughness and in certain cases one of the faces was additionally polished. Studies were performed under natural sun light and the solar zenith angle varied between 46° and 65°.

BRIEF PETROGRAPHIC CHARACTERISTIC OF THE ROCKS

Samples of which the SRCO was measured were taken from various outcrops observed in different parts of Bulgaria and are selected in such a way as to be representative of magmatic and metamorphic rocks.

The volcanic rocks are taken from Rhodopes and the Western part of Sredna gora mountain. They relate to acid and medium acid volcanites. 11 samples are examined here: rhyolites (No. 1-6, rhyodacite (No. 7), dellenites (No. 8-9), latite (No. 10) and andesite (No. 11). They occur as flows, subvolcanic bodies and dikes.

The granitoides are taken from Rilo-Rhodopean batholites. Here are granitoides Nos. 3011, 4732, 4584, biotite granites Nos. 2086, 2193, 224, 3735, 5368, 8653 and two-mica granites Nos. I-II.

Two types of gneisses are examined: biotite and two-mica (No 340, 1907, 5005).

DISCUSSION OF RESULTS

Both within the period of measurements, and when discussing the results, factors with greatest influence on rock SRCO can be defined separately into two principal groups: internal and external.

Between external factors the main role is played by the source of light which may be of natural or artificial origin, but its influence is known as a rule and could be considered in data interpretation.

Between internal factors significant are structural-textural features, mineral and chemical composition, and sample surface roughness.

When measuring the spectral brightness of the different genetic classes of rocks, sun was used as a source of light. The strength of solar irradiation for the geographic latitude of Sofia (Eastern longitude - 1h 33m 23sec; Northern latitude - 42° 41' 02') depends on solar location over horizon, i.e. on the zenith angle. Measurements were performed during the autumn of 1978 on Sept. 30 and Dec. 3-4, within the interval of 11:23-15:51. During these days the zenith angle has changed from 46° to 65°. When comparing the changes of

SRCO measured values for the three days and for the interval of noon hours during autumn, it was determined that the solar zenith angle was of little importance.

Volcanites. Data obtained in sample measurements of acid volcanic rocks show that the highest is the SRCO of sample No. 1 - rhyolite, taken from the Eastern Rhodopes. Sample No. 2 - rhyolite from the same volcanic region is defined by lowest indices of spectral brightness. Both samples do not differ in such structural indices as shape, size of porphyries, but there is considerable difference in the quantity of ground mass. Referring to this index, the acid volcanic rocks may be divided into two groups: 1) with great quantity of ground mass (over 80%) for samples Nos. 1, 6, 7 and 2) with higher porphyric content (30-40%) for samples Nos. 2, 3, 4, shown in Fig. 1. The rock samples of the first group are characterized by high values of SRCO within the range of 600-800 nm. Sample No. 5 - rhyolite under equal porphyries and ground mass has average indices of reflective capacity with respect to the upper two groups (Fig. 1). Sample No. 7 - plagiorhyolite-rhyodacite has the following specifics: within the range of 400-500 nm the spectral brightness equals the rhyolitic one, but within the range of 600-800 nm it increases, which can be interpreted with the structural features of the rock. It is of dense almost aphanitic structure for the porphyry and holocrystalline for the ground mass. This structural index of the sample brings it closer to the rhyolites with higher amount of ground mass. The small mineral sizes definitely influence the SRCO increase within the range of 600-800 nm. In augmenting the porphyritic sizes, the effect on the spectral brightness is reverse. Typical example is the coarse grained porphyritic dellenite (sample No. 9). Here the lowest values of spectral brightness are observed - 10% within the short waves interval of the electromagnetic spectrum up to 24% in the IR range. The other sample (No. 8) of coarse grained porphyritic dellenite does not differ from No. 9 in structural indices (size and shape of minerals) but its indices are an average of 10% higher. The macroscopic comparison of the two samples shows a difference between the surfaces, on which measurements are taken. For sample No. 9 it is rough and uneven with a difference in roughness up to 5 nm. At other equal conditions one more index is introduced and namely the surface of the measured samples. Under natural conditions this index probably would be of lesser importance because of the physico-chemical rock weathering, dust, and development of micro-organisms.

The dominant colour of the acid volcanic rocks is rose-violet and pale rose. Sample No. 6 - perlite represents an exception. It is almost white and sample No. 4 - rhyolite is of intensive red colour. In fact only these rock samples have shown SRCO indices, which differ from the others. This effect is especially remarkable within the short waves interval of the electromagnetic spectrum, between 400-500 nm. For this range the perlite shows values of spectral brightness by 10% higher and with reference to the rhyolite of intensive red colour the difference attains up to 15%.

Sample No. 10 - latite and No. 11 - andesite refer to the medium acid rocks. Here a steep decrease of the reflective capacity is observed. A slow increase of the SRCO is found for the latite - from 14% to 18%. The lowest values of spectral brightness - from 10.5% to 17.5% are measured for the andesite. Obviously the dark colour minerals increase significantly decreases the reflective capacity of the volcanic rocks and this decrease is to be traced through the whole measurement range of the electromagnetic spectrum. Referring to the colour of the medium acid rocks: grey-greenish for the latite and dark grey for the andesite, the following peculiarity was observed - for all measurements the SRCO of the latite is by 4-5% higher as compared to the andesite.

Measurements have shown that the rock surface influences significantly the SRCO and that is why one of the sample faces was additionally polished (Fig. 1). It was found that within the interval of 400 nm to 550 nm the SRCO increases by 10% at average and to the near infrared range gradually decreases and is of about 5%.

Granitoides. When the SRCO of granitoides was measured it was determined that its highest values were attained with the two-mica granites followed by gradual decrease to the biotite granites and the granodiorites (Fig. 2). The two-mica granites SRCO changes from 48-51% at 402 nm to 59-64% at the near infrared region. The lower values of sample I compared to sample II are due to bigger mineral grain size. The biotite granites have SRCO from 28% to 60%. The average values are between 35% and 50% (Fig. 2). Lowest are the values of sample No. 2224 (from 28% to 40.5%). We assume that this is due both to the weathered sample surface and greater quantities of biotite. Sample No. 2086 has the highest SRCO values - 62% at 775 nm. This increase results from the augmented content of salic minerals as the percentage of potassium feldspar is particularly high.

Results from the biotite granodiorites SRCO measurements show that they have lowest values related to the other granitoides. An exception here is sample No. 4584, for which the SRCO changes from 45% to 55%, while for all the other samples it is between 22% and 35%. Samples do not differ in structural features and mineral composition. The difference appears in the gravity ratio between mafic and salic minerals. Sample No. 4584 contains about 65% light minerals (quartz, feldspar, plagioclase). Sample No. 3011 has lowest values - here the mafites are predominant.

Gneisses. They exhibit lowest values of SRCO. For examples Nos. 1907, 5005 (biotite gneisses) it changes from 13-20% at 402 nm to 30-31% at 802 nm. The lowest SRCO was surprisingly determined for the two-mica gneiss No. 304 (from 7.5% to 27%). We assume that the reflective capacity decrease is due to well expressed banded texture.

CONCLUSIONS

After the measurements of the representative samples of the two genetic rock classes, the following conclusions can be made:

- The presence of two plateaus is a typical peculiarity for the rock formations, when deducing the SRCO. The first is defined in the short wave length interval between 400-500 nm with a very poor decrease within the interval 411-425 nm. The second plateau is within the range of 700-802 nm also with a weak decrease after 775 nm. In fact the two plateaus represent ranges within the optic spectrum of up to 100 nm wide, where constant reflective capacity is observed.

- Another clearly expressed peculiarity of the rocks is the SRCO rapid increase within the interval of 500-700 nm.

- Acid volcanic rocks express direct proportional dependence between ground mass quantity and SRCO.

- Acid rocks have higher SRCO values than the medium acid or basic rocks.

- The predominant quantity of salic minerals augments the SRCO and vice versa.

- The rock structure effect is manifested through the size of the minerals: at greater sizes the SRCO decreases and the reverse.

- Smooth and polished rock surfaces increase the SRCO.

- Samples of weathered surface have lower SRCO than the fresh ones.

- Rocks of massive texture have higher SRCO than the ones of parallel (banded) structure.

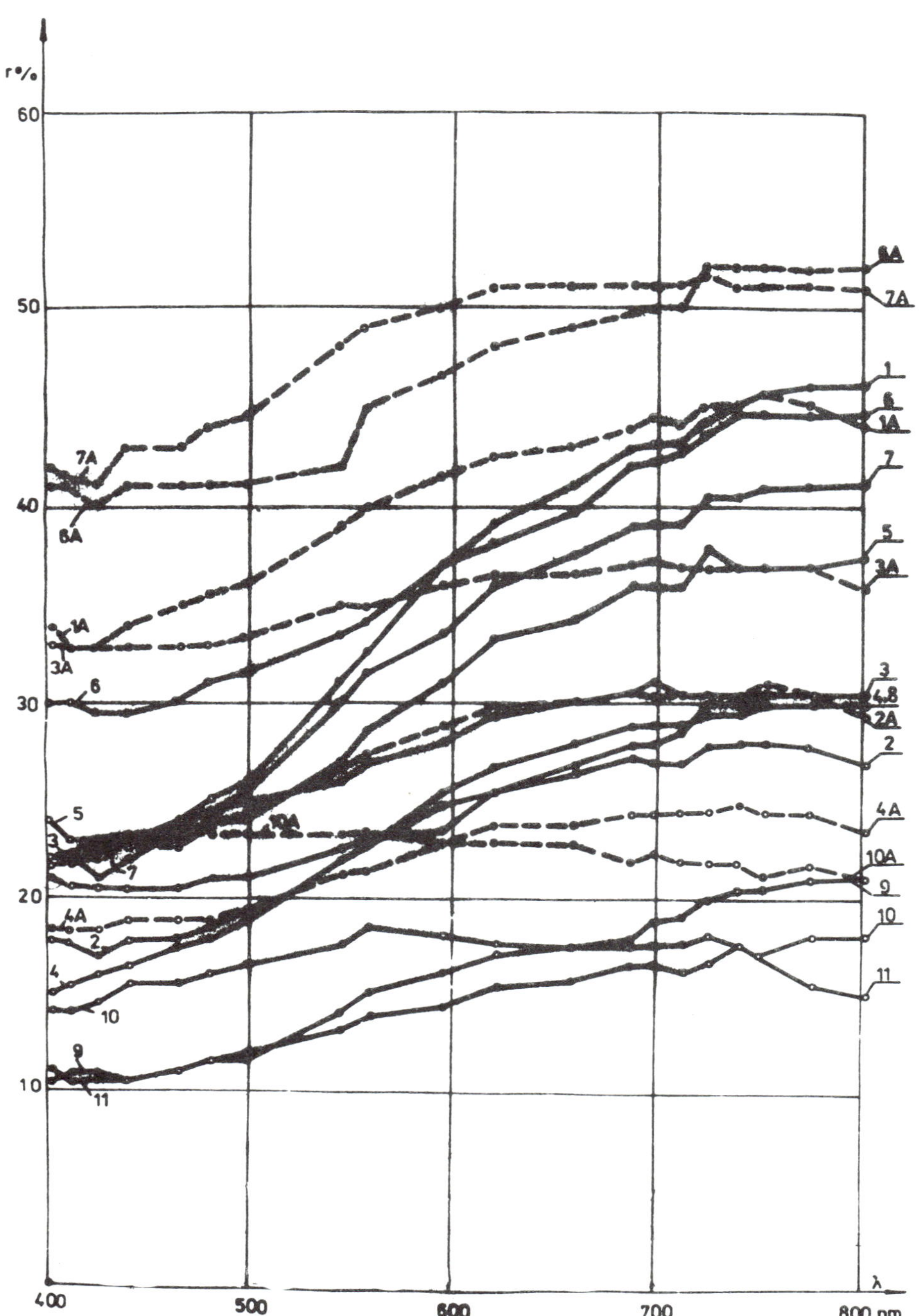

Fig. 1 - SRCO curves of volcanites. Samples: rhyolites (Nos. 1, 2, 3, 4, 5); perlite (No. 6); plagiorhyolite-rhyodacite (No. 7); dellenites (Nos. 8, 9); latite (No. 10); andesite (No. 11). Broken curves refer to polished samples.

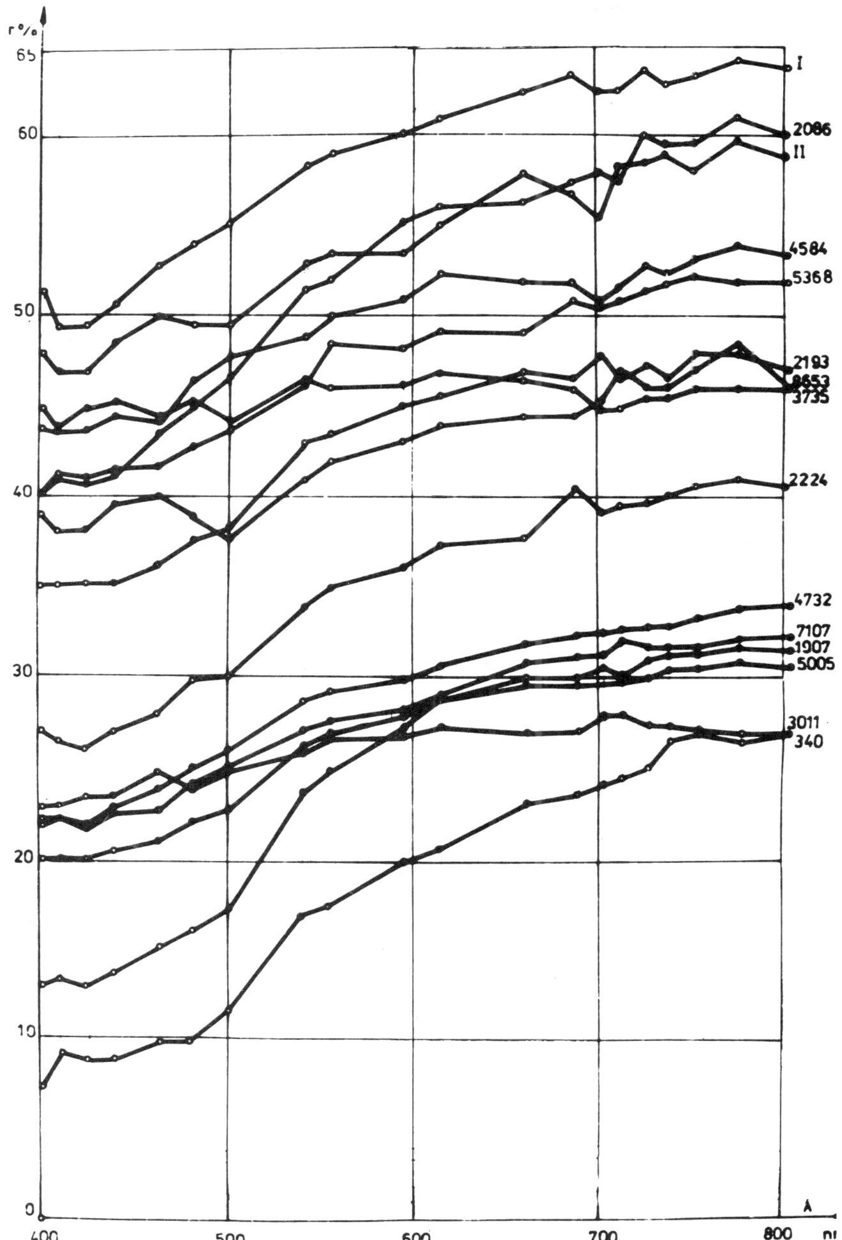

Fig. 2 - SRCO curves of granitoides and gneisses. Samples: biotite granites (Nos. 2086, 5368, 2193, 3735, 8653, 2224); two-mica granites (I, II); biotite granodiorites (Nos. 4584, 4732, 3011, 7107); gneisses (Nos. 1907, 5005, 304).

AN AUTOMATIC METHOD OF DISCRIMINATING ROCK OUTCROPS USING LANDSAT DATA

P. Chagarlamudi

The Sibbald Group, a division of Deloitte, Haskins & Sells Associates, Ottawa, Canada

ABSTRACT

A method of discriminating rock outcrops has been developed using Landsat data. The method, based on a hierarchical classification scheme, has been calibrated and tested in the Coppermine River area of the Northwest Territories of Canada. Maps showing outcrop locations have been prepared using this method. These maps were compared with the published geological maps, field notes, and aerial photos. The study shows that rock outcrops can be uniquely identified and that discrimination among different rock types is possible. Further, the results indicate that the Landsat data are as good as aerial photos for locating rock outcrops.

INTRODUCTION

Geologists spend large amounts of time in preparing maps prior to launching a field mapping program. These pre-field maps are usually prepared with the aid of aerial photos when they are available. These maps are a valuable aid in planning and in carrying out the mapping. Satellites may provide an alternative data source for making these maps. The Multi-Spectral Scanner (MSS) data from the Landsat satellites are available in four spectral bands referred to here as Bands 4, 5, 6, and 7. The MSS on Landsats 1 and 2 has six detectors sensing 64 grey levels for each spectral band [1]. The MSS data are most useful in digital form.

The objective of this study was to develop a computer method of discriminating rock outcrops using Landsat data. The digital data were standardized so that rock outcrops were automatically separated without the use of training data. The method was tested to determine whether the maps generated could provide an alternative to those produced from aerial photos.

DATA AND METHODS

The Study Area

The study area is located just south of Coppermine, Northwest Territories, Canada. The advantages of this area are that rock exposure is abundant, the outcrops are relatively free from tree cover and surficial material, different rock types are present, and detailed geological maps and field notes exist for verification. Generalized geology of the study area, superimposed on an aerial photo, is shown in Figure 1.

Standardizing Landsat Data

Collecting ground data for use in calibrating Landsat data is expensive. Repeated calibrations may be eliminated by using standardized Landsat data and threshold values. The standardized calibration data used in this study were defined for all future applications of the method during a pre-calibration study. The MSS data used were linearized and expanded to 255 grey levels from the original 64. These data were then standardized for differences between the detectors within each band, between the satellites, in the sun elevation, and in the atmosphere. The differences in grey levels still present in the data represent the physical variations on the ground. This standardization process has been used to compare Landsat data acquired on different dates in a water sediment-level monitoring system [2].

The Classification Scheme

There are no unique four-band signatures in the MSS data for the rock classes in the study area. Because of this, conventional classification procedures, which employ all the four bands simultaneously, have not been useful for separating rock classes [3]. It was discovered in this study that there were, however, distinct patterns for rock classes. A hierarchical classification scheme was developed to recognize these patterns. For this, individual bands and arithmatic combination of bands were selectively employed. In the hierarchy, rock outcrops are first separated from all other classes, then the outcrops are tentatively identified as rock classes.

In the first step, a previously described hierarchical separation scheme was used [4]. In this scheme, threshold values in Bands 7, 5 and the ratio of Band 7 to Band 5 were used in succession to eliminate water, clouds, and vegetation. In addition, a Band 4 threshold value was used to eliminate snow and ice (Fig. 2). The remaining area is classified as rock outcrop and surficial material.

For the second step, the threshold values for the best bands and band combinations for separating the four rock classes present in the study area were selected. These selections were made using averages of standardized MSS data for 36 pixels in calibration areas for each class. The bands and band combinations selected are shown in Figure 2.

Producing the Rock Outcrop Map

To be useful in the field, any rock outcrop map produced must directly overlay on a standard map projection. To accomplish this, an affine transformation and resampling with a cubic convolution algorithm were used to map the data into the Universal Transverse Mercator (UTM) grid system. The Landsat data are classified by pixel area (50 X 50 m), assigned symbols and mapped on a computer print-out at 1:60,000 scale. Mensuration data on all classes are also produced automatically for estimating the time requirements of the field studies.

RESULTS AND DISCUSSION

Exposed rocks in the study area were, in general, accurately identified by the automatic method. A map of rock outcrops in the study area produced from the MSS data acquired on 5 August 1975 is shown in Figure 3. Two areas of rock outcrops, marked R, are shown on both the outcrop map (Fig. 3) and the aerial photo (Fig. 1) for comparison. Areas of glacial drift on the geological map, including such surficial materials as sand, soils and boulders, were often identified as vegetation on the outcrop map. Glacial drift areas near the lakes and streams were, however, identified as surficial materials on the outcrop map.

Discrimination among the different rock types was also accurate in the area. Outcrops identified by their respective rock type in the second step of the classification scheme are shown on the outcrop map (Fig. 3). This map was compared with the aerial photo and known geology (Fig. 1). On the photo, more outcrops appear in sandstone and dolomite areas than in areas of granites and basaltic flows. Most of the sandstone and dolomite outcrop areas were correctly identified on the map. As an example, a large U-shaped exposure in the sandstones is marked on Figures 1 and 3 for comparison. Most areas of basalts and granites were covered with vegetation, as seen on the photo, and these areas were identified as vegetation on the outcrop map.

The same threshold values were also tested using Landsat data acquired on 22 June 1975. The identification of outcrop locations was accurate but the discrimination among the rock types was poor. The subtle differences present among the rock types in the area were probably masked by the presence of snow and ice.

The cost of producing the outcrop maps using Landsat data would be less than using aerial photos. The estimated cost of producing an outcrop map of a 100 x 100 Km area on a computer is $50 and the cost of geometrically corrected current Landsat data is $110. It is estimated that three-man days, at $250/day, would be needed to prepare the map from existing aerial photos (sixty at 1:60,000 scale) costing $90. Thus, producing the map from current Landsat data would cost $160 compared to $840 from aerial photos. The cost of acquiring aerial photos would greatly increase the cost difference.

CONCLUSIONS

The distribution of rock outcrops on maps obtained using the automatic method with Landsat data developed in this study agrees well with known geology and aerial photos. Rock outcrops were consistently discriminated on two different dates using standardized Landsat data. Discrimination on August data among different rock types was also accurate. The cost of generating outcrop maps from Landsat data by the automatic method developed is estimated to be less than 20% of the cost of generating the maps with aerial photos using conventional methods.

ACKNOWLEDGMENTS

Discussions on the geology of the area with Drs. J.A. Donaldson, W.R.A. Baragar, P. Hoffman and Mr. H. D. Moore have been very helpful. Sincere thanks are to Drs. J.S. Schubert, A.F. Gregory and D.H. Hall for helpful discussions.

REFERENCES

1. NASA, Landsat Data Users Handbook, Goddard Space Flight Center, Greenbelt, Maryland, Document No. 76SDS4258, 1976.

2. P. Chagarlamudi, R.E. Hecky, and J.S. Schubert, Quantitative Monitoring of Sediment Levels in Freshwater Lakes from Landsat, this volume.

3. B.S. Siegal, and M.J. Abrams, Geologic mapping using Landsat data, Photogrammetric Engineering and Remote Sensing XL11, 325 (1976).

4. J. Schubert, P. Chagarlamudi, H. Moore, and C. Goodfellow, Globial Agricultural Productivity Estimation Project, Gregory Geoscience Ltd., Report to Agriculture Canada under contract # 1SW5-0368, 1977.

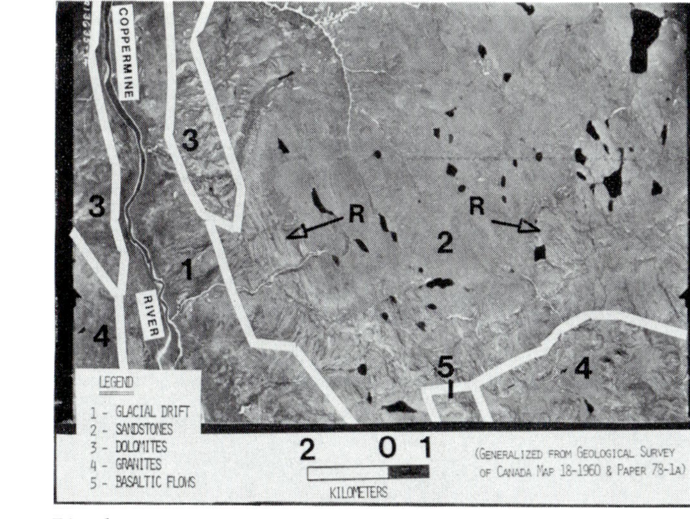

Fig.1 Known geology superimposed on an aerial photo.

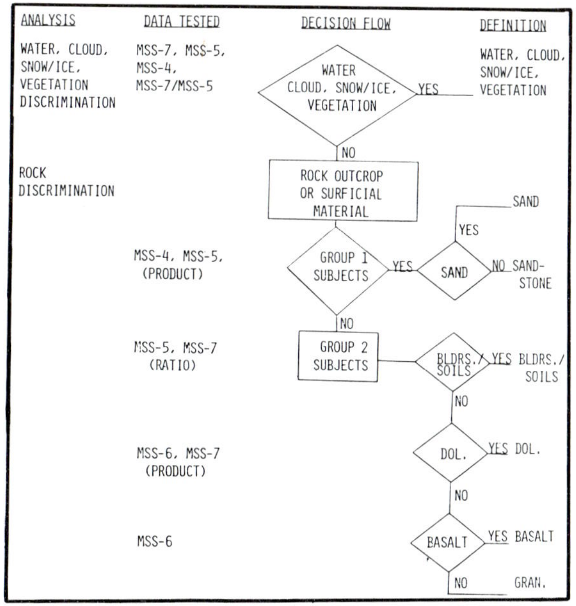

ANALYSIS | DATA TESTED | DECISION FLOW | DEFINITION

WATER, CLOUD, SNOW/ICE, VEGETATION DISCRIMINATION — MSS-7, MSS-5, MSS-4, MSS-7/MSS-5 — WATER CLOUD, SNOW/ICE, VEGETATION — YES — WATER, CLOUD, SNOW/ICE, VEGETATION
NO
ROCK DISCRIMINATION — ROCK OUTCROP OR SURFICIAL MATERIAL
MSS-4, MSS-5, (PRODUCT) — GROUP 1 SUBJECTS — YES — SAND — YES — SAND; NO — SAND-STONE
NO
MSS-5, MSS-7 (RATIO) — GROUP 2 SUBJECTS — BLDRS./SOILS — YES — BLDRS./SOILS
NO
MSS-6, MSS-7 (PRODUCT) — DOL. — YES — DOL.
NO
MSS-6 — BASALT — YES — BASALT
NO — GRAN.

Fig.2 The decision flow sequence in the hierarchical classification scheme developed.

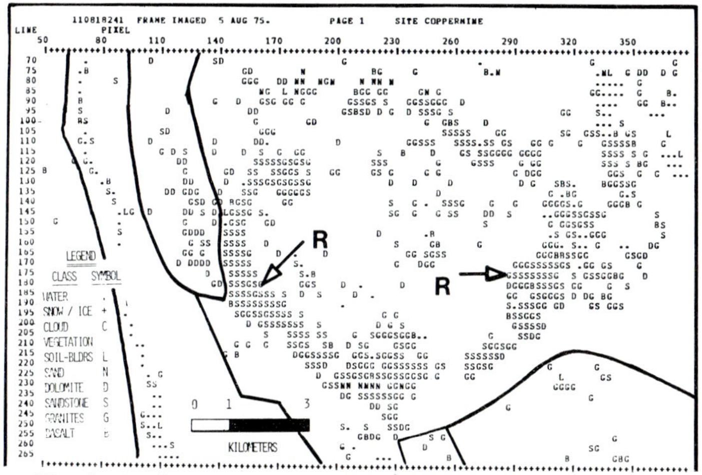

110818241 FRAME IMAGED 5 AUG 75. PAGE 1 SITE COPPERMINE
LINE PIXEL 50 80 110 140 170 200 230 260 290 320 350

LEGEND
CLASS SYMBOL
WATER .
SNOW / ICE +
CLOUD C
VEGETATION ,
SOIL-BLDRS L
SAND N
DOLOMITE D
SANDSTONE S
GRANITES G
BASALT B

0 1 3 KILOMETERS

Fig.3 Rock outcrop map produced of the study area.

L-BAND RADAR AND GEOLOGY: SOME RESULTS IN SOUTH-EAST OF FRANCE

P. Rebillard

Institut Dolomieu-Géologie et Minéralogie, Rue Maurice Gignoux, 38031 Grenoble, France

ABSTRACT

One L-Band Side Looking Airborne Radar experiment was carried out in the S.W. of the French Alps in June, 1977. A comparison has been established between geological maps and aerial views taken with HH and HV polarization. Two results are remarkable : - The roughness of the objects can influence the image on the pictures in HH polarization,

- White marks in HH (and black in HV) can be seen, a field analysis has shown the structural reality of this line.

These results should be correlated with results obtained from others sensors as SEASAT.

INTRODUCTION

This paper presents the first results of an interpretation of Side Looking Airborne Radar (SLAR) pictures taken in the South-East of France in 1977. This experiment was carried out in the night of the 22nd-23rd June '77 with a L-Band Radar from the Jet Propulsion Laboratory. This sensor was placed on board of a NASA's Convair 990, at the altitude of 12 000 meters. Fourteen flight paths were chosen by the "Groupement pour le Développement de la Télédétection Aérospatiale".
The flight path number 3 of these experimental datas, over two areas in the southern Alps at the scale of 1/500 000, was interpreted by the Geological Laboratory of the Univ. of Grenoble (France) jointly with the "Laboratoire d'Applications Spéciales de la Physique" at the "Centre d'Etudes Nucléaires de Grenoble".

FLIGHT LOCATION AND GEOLOGICAL PRESENTATION

On figure 1, we can see the two flight paths of the radar coverage :
- the 1st one from Allos to the Rhône valley, in the southern area (3a, 3b),
- the 2nd one from Valdrôme to Montélimar for the northern area (3c, 3d).

On the structural map (Fig. 2) we can define the two areas of our research :
- Nyons-Condorcet in the West,
- Laragne-Sisteron in the East.

Using classical methods of the photo-interpretation, a comparison has been established between geological maps and aerial views taken with HH and HV polarization and enlarged at the scale of 1/100 000. Because the radar was pointed North, the presented data show the North and the South in inversed positions. Therefore, the

Fig.1 Flight location and radar characteristics.

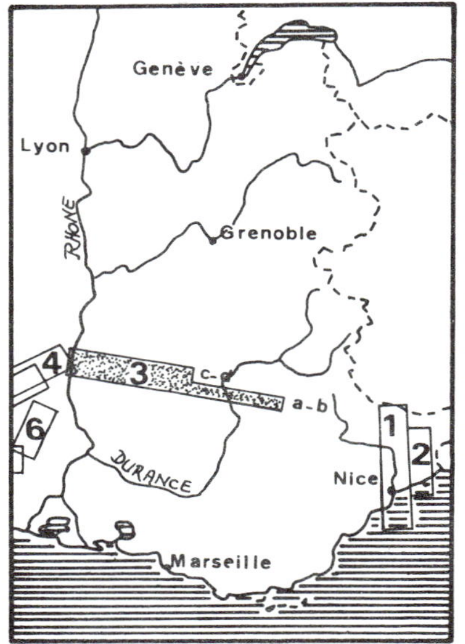

Wavelength	:	25 cm (L band)
Power	:	4 kW
Antenna	:	75 × 25 cm
Polarization	:	HH and HV
Resolution	:	longitudinal 25 m transverse 20 m
Ground band coverage	:	15 km
Scale	:	1/500 000
Time	:	23.07 to 23.20 G.M.T.

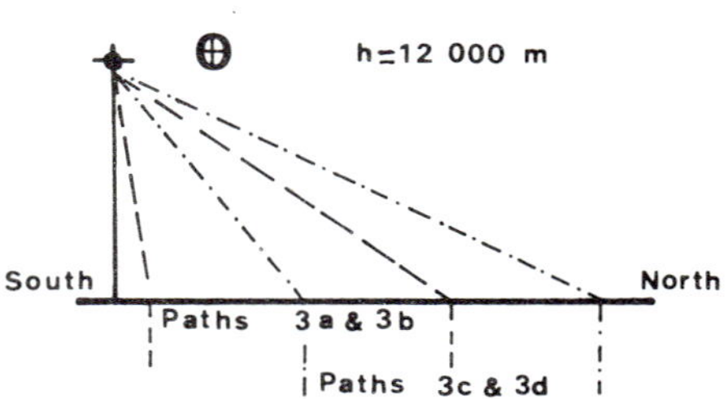

Fig.2 Structural map of south Diois and Baronnies, location of the areas studied.

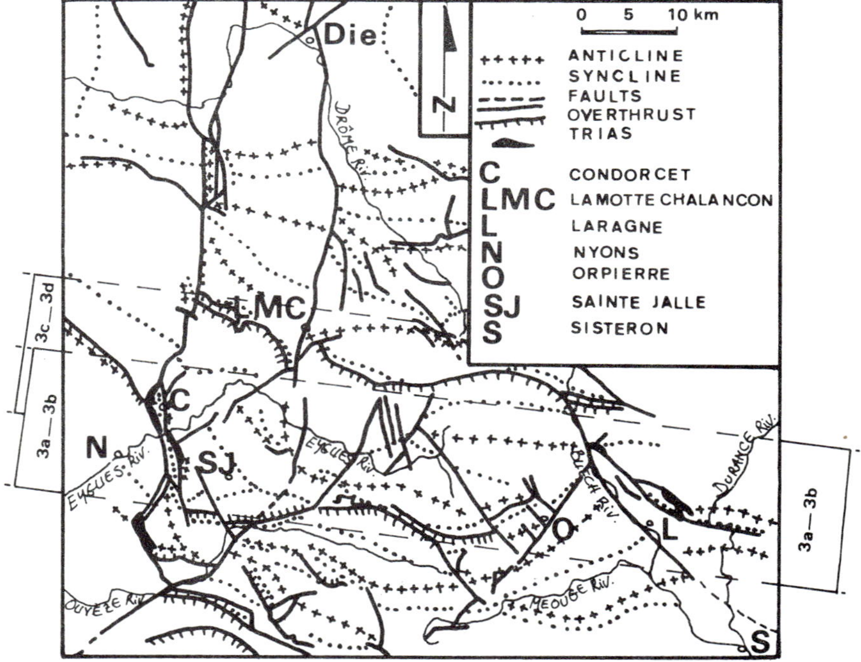

crest of mountains represents the horizontal trajectory of the plane.

The overflown area belong to the subalpine southern chains and is characterized by N-S, NE-SW and NW-SE structural features [2 and 4]. Along these a triassic diapir emerged near Condorcet and Montaulieu. The stratigraphical succession is continuous from the middle Jurassic to the upper Cretaceous.

The Rhône valley is composed of Cainozoic sandstone deposits (Miocene and Pliocene). The Oligocene is absent and the Eocene exists South of Nyons.
These two zones were chosen for their metalliferous significance. Indeed, at Condorcet from 1904 to 1940 many companies extracted such minerals as Barytes, Galena and Strontium in contact with the Triassic diapir. At Orpierre the mining works ended in 1900. This particular mining operation concentrated on Iron which is to be found in Calcite veins from NE 10° to NE 15°.
In the interpretation we encontered some difficulties due to the height of the mountains [3]. However it is possible to see the hydrological, structural and geomorphological characteristics.

1. Hydrological observations (Fig.3)

On these photographs we can see the Durance river (1) in the East and the Buëch river (2) in the West. The two rivers apear differently.
With HH or HV polarization the Durance bed is clearly seen : it is possible to obtain a map of the major bed (1a), the minor bed (1b) and the ancient beds (1c). For the Buëch river these details can only be observed with HV polarization, the HH polarization do not give any results.
One explanation of this difference lies in the roughness introduced by the river flow, another one is the nature of the alluviums or by the sediments carried by the flow.

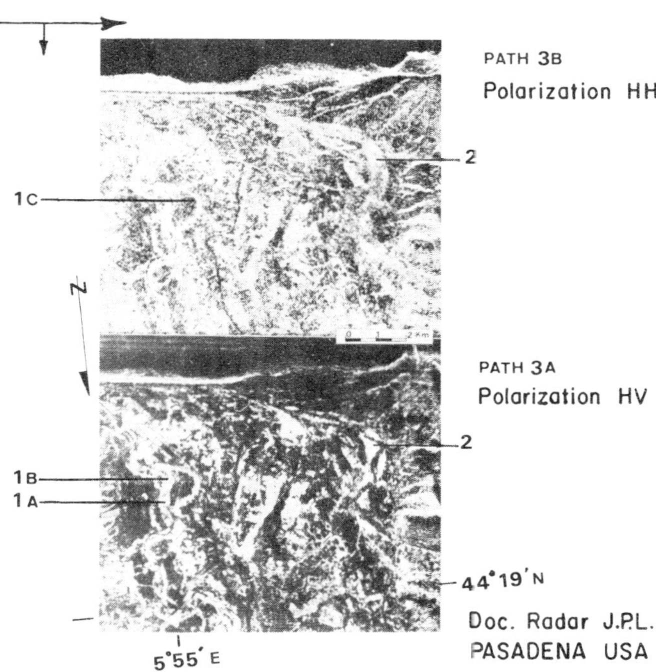

2. Structural observations (Fig.4) At the Soubeyrand Pass (1) many faults can be seen and they are related to the big submeridian fracture running between the region of Die and La Motte Chalancon (Fig.2).
Another fault affects an anticline in a dextral shearing manner (2). Many other faults (3) trending North-South were described from the Pass to the Eygues river which appear NE-SW in the pictures. This is due to the distortion of the pictures in this mountainous area.
On the left side of the Ennuye river (4) a white linear mark on the HH view can be observed although it appears black on the HV picture. The geological map and the aerial photographs do not show any fault.
A ground work has shown the structural origin of this mark, which corresponds to a succession of little normal faults oriented N150, 70E. The observations were difficult because of the mediterranean vegetation.
A number of propositions can be drawn to explain the presence of these minor geological features on the pictures and their different behaviours in HH and HV polarization.

Some lithological boundaries can also be observed as at Condorcet and in the Rhône valley (Fig.4).

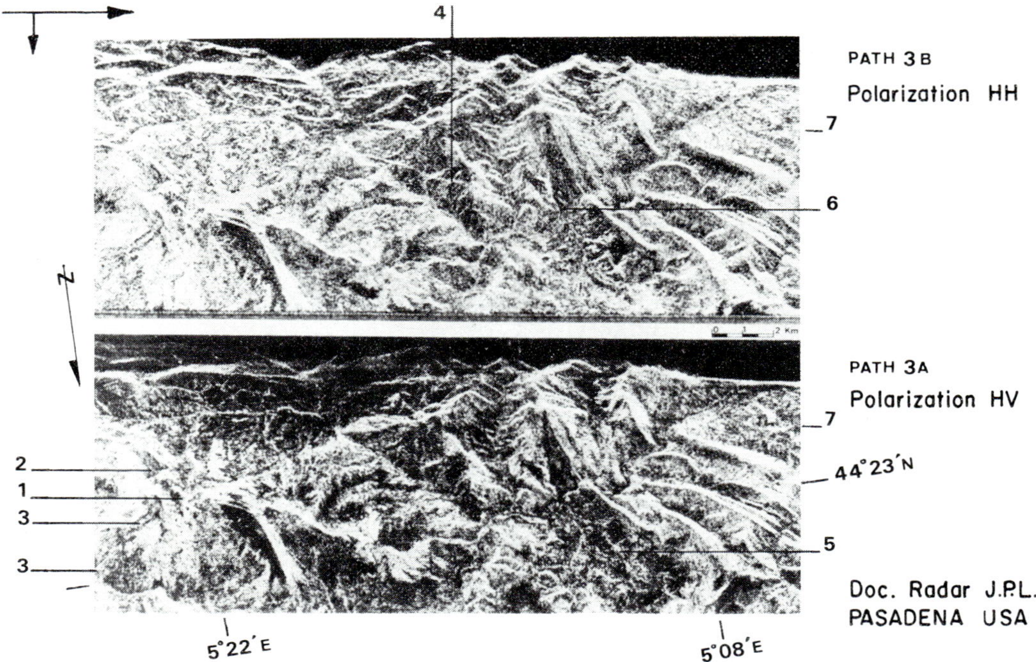

At Condorcet, the triassic diapir (5) can only be seen with the HV polarization and not in HH due to it roughness. The diapir of Montaulieu (6) does not appear because of it situation on the northern side of a hill.
In the Rhône valley we can define the boundaries of the Pliocene (7). But this observation is only possible with the vegetation, indeed these hills are covered with boxwood, oaks and thorn bushes and vineyards are planted on the Miocene.

CONCLUSIONS

The applications of the L-Band radar are mainly on structure and hydrology maping, but the distortion of the pictures and the resolution prevent a good topographical location. These new results applied on an alpine area should be correlated with results obtained from Seasat or new JPL air-missions.

This work was supported by the Centre National d'Etudes Spatiales, jointly with the Institut Dolomieu and the Laboratory A.S.P.

References

1. DEANE, R.A. Side looking radar systems and their potential applications to the earth ressources surveys. E.S.R.O., 136, 137, 1973.
2. FLANDRIN, J. Sur l'âge des principaux traits structuraux du Diois et des Baronnies. Bull. Soc. Géol. de France (7) VIII, 376-86, 1966.
3. REBILLARD, Ph. Programme radar JPL 1977. Analyse des clichés de l'axe de vol n° 3. Diois-Baronnies (France). CENG/ASP 79-03, 1979.
4. VIALON, P. et al. L'arc alpin occidental : réorientation des structures primitivement E.W. par glissement et étirement dans un système de compression N.S. ? Ecl. Geol. Helv. 69, 2, 509-19, 1976.

AUTHOR INDEX